Ethnographies of Prostitution in Contemporary China

ETHNOGRAPHIES OF PROSTITUTION IN CONTEMPORARY CHINA

GENDER RELATIONS, HIV/AIDS, AND NATIONALISM

Tiantian Zheng

First published in 2009 by PALGRAVE MACMILLAN® in the United States—a division of St. Martin's Press LLC, 175 Fifth Avenue, New York, NY 10010.

Where this book is distributed in the UK, Europe and the rest of the world, this is by Palgrave Macmillan, a division of Macmillan Publishers Limited, registered in England, company number 785998, of Houndmills, Basingstoke, Hampshire RG21 6XS.

Palgrave Macmillan is the global academic imprint of the above companies and has companies and representatives throughout the world.

Palgrave® and Macmillan® are registered trademarks in the United States, the United Kingdom, Europe and other countries.

ISBN: 978-0-230-61741-4

Library of Congress Cataloging-in-Publication Data is available from the Library of Congress.

A catalogue record of the book is available from the British Library.

Design by Scribe, Inc.

First edition: July 2009

10 9 8 7 6 5 4 3 2 1

Printed in the United States of America.

I dedicate this book to

Helen Siu   *Harold Scheffler*

Deborah Davis   *William Kelly*

CONTENTS

Acknowledgments

I would like to extend my sincere thanks to all the women and men in Dalian who shared with me their experiences, thoughts, and feelings about sexuality, contraception, condom uses, and sexually transmitted diseases and HIV/AIDS. Without their support and contribution, fieldwork for this book would have been impossible.

Over the years of my research and writing on karaoke bar hostesses and clients in Chinese sex industry, I accumulated a great deal of debt to my professors, colleagues, and friends. I want to extend my special thanks to Evelyne Micollier. I benefited from her generous, thorough, and incisive comments that have sharpened the arguments and improved the manuscript. I thank her for the time and energy she spent in reading the manuscript. I thank her for the careful work with the book. I thank her for the helpful and constructive suggestions. I thank her for the warm encouragement and support. Her creative and keen comments have helped me enhance the quality of the book.

I would like to thank my advisors at Yale University, Deborah Davis, William W. Kelly, Harold W. Scheffler, and Helen Siu, for their continuous and incessant support of my professional development after I graduated. I would also like to thank Susan Brownell for her persistent care and support of my professional development. Through setbacks, they have always showed me great empathy, support, care, and faith, and offered me helpful professional advice. I thank them for having taught me, supported me, and encouraged me more than they could realize, for which I am forever grateful.

Hal Scheffler was especially enthusiastic and interested in this research. He urged me to e-mail him my writing so that he could provide feedback and comments. I e-mailed him my book proposal and a book chapter, and within a week, I received from him an edited copy full of his helpful and insightful comments. He continued to be an intellectual inspiration for me, as he continued discussing with me about my new research and e-mailing me new information about books, articles, and research that would be helpful for my research. He also offered me excellent advice on the choice of the book publisher. His endeavor to support me intellectually and professionally moved me deeply. I thank him for his continuing support.

I would also like to thank my professor, mentor, and friend, Jack Wortman, for being a constant intellectual and emotional support for me during the past fourteen years. I thank him for his consistent mentorship in every aspect of my life—emotional, professional, and academic. I thank him for always being there for me whenever I need someone to talk to. I thank him for being a patient and faithful listener. I thank him for providing me helpful and wise advice on problems and issues in my life, whether emotional, professional, or academic. I thank him for being a selfless and generous advisor who offered me an immense amount of time in helping me correct the language of every article and every chapter I wrote. I also thank him for engaging in inspiring and insightful discussions with me on the materials. My debts to him and to all the above-mentioned professors can never be repaid.

My thanks also go to those who have offered helpful comments on parts of the book, including Mark Selden, and other anonymous reviewers for parts of the book. I would like to thank William Kelly for inviting me to Yale to present parts of the book. Thank you to Erik Bitterbaum, Mark Prus, Amy Henderson-Harr, and Glen Clark for offering me conference grants to present parts of the book at conferences. Thank you to Amy Henderson-Harr and Glen Clark for their generous help with the grants to conduct the fieldwork for the research.

My special gratitude also goes to my parents who were encouraging and helpful during my fieldwork. Thank you to Dave Grass for his emotional support and positive encouragement, which have made the writing process enjoyable and pleasant. Thank you to Brigitte Shull for her enthusiasm about my project and for her support in guiding me through the book preparation and production process.

INTRODUCTION

IT WAS 10 P.M. ON A FRIDAY night. A night show was on, punctual as usual, in the dimly lit huge rest hall at a renowned Dalian sauna bar. Around one hundred male customers who had finished their sauna bath slouched on the cushioned sofas, in sleeping gowns provided by the sauna bar. They were drinking soft drinks, having their feet massaged by service workers, while watching the show. Female hostesses, in their early twenties, dressed in revealing clothes, glided from one aisle to another, murmuring in each customer's ear, asking if he needed any "special service."

I was sitting next to the three male customers who were participants in my research. They had brought me to this night show so that I could "experience the culture of male customers in the sex industry." A hostess approached one of them and asked, "Would you like a Korean-style bone-loosening massage?" He pretended he did not understand, "What is it for?" Smiling, the hostess answered with a coquettish yet seductive tone, "Well, this service is the kind that can satisfy you; the kind that can get your problem resolved!"

The night show continued with dancing, singing, and sexually explicit and crude comedies. Between performances, the emcee urged the audience to participate in the lottery game—20 yuan for each lot for an award of 500 yuan. The emcee repeatedly announced to the audience that this was a chance to test their luck. He yelled, "Let's just play!" He continued, "Play is happy [*kaixin*], play is complete [*tongkuai*], play is absolute [*xiaosa*], play is stimulation [*ciji*]. Let's sell our children and buy a monkey to play with! Let's dismantle our houses and look for a cricket to play with! Let's just play!" He repeated these words over and over for almost an hour. Under his influence and the influence of the beautiful hostesses who egged on each customer, one after another shouted out loud, "I want one." "I want two." "I want three." Each of the three male customers sitting next to me bought lotteries. Each departed to a special "rented room" with a hostess to engage in the special service.

I witnessed this scene during my fieldwork in 2005. It conveys an outright rejection of mainstream morality and an absolute embrace of hedonistic pleasure, free from moral constraints. The hyperbolic chatter is meant to be humorous. Yet it embodies the hedonistic mood that dominates clients' attitudes. In this milieu, clients are urged to reject what society tells them is important and reject mainstream values. That is, embrace extreme and absolute pleasure and renounce everything imprinted or tinged with state- and family-inculcated values. Condoms fall into this category, symbolizing state repression and control of sexuality. Clients regard the dissemination of AIDS information as just another state strategy to scare them and control sexual behavior. Hence, in their group, they tend to reject condoms.

This book draws on my two years of fieldwork in Dalian between 1998 and 2002 and twelve months of fieldwork in Dalian between 2004 and 2007. On the basis of three years of extensive fieldwork, this ethnographic study of prostitution in the metropolitan city of Dalian, China, explores the drama of survival as rural migrant women working as karaoke bar hostesses struggle to cope both physically and emotionally with the dangerous world of the karaoke bar. It delves into the interplay of gender politics, nationalism, and power relationships that inhere in practices of birth control, disease control, and control of women's bodies. It provides an ethnographically rich account of how karaoke bar hostesses and clients handle HIV risks; how they perceive condoms; how they define and conceptualize sex, sexual desires, and different sexual practices; and how they make decisions about birth control and condom use during illicit sex.

More specifically, the book will examine the following questions: What do clients and hostesses know about the risks of unprotected sex and the benefits of condom use for themselves? Where does their information come from and how accurate is it? How do clients and hostesses perceive condoms and how do these perceptions relate to (non-)condom use? Through addressing these questions, this book will demonstrate the contribution of ethnography to the study of public health, gender, and sexuality.

HIV/AIDS IN CHINA

UNAIDS (2006) estimates that less than 0.1 percent of the total population of China is HIV positive. This number may seem small when compared with many other nations where, in some populations, the comparable figure is as high as 30 percent. But all signs are that the numbers are increasing rapidly. Over the past two decades, China's AIDS epidemic has progressed from a localized epidemic of injection drug users (IDUs) along the border with Myanmar in Yunnan province to thirty-one provinces, municipalities, and autonomous regions (Kaufman, Kleinman, and

Saich 2006b, 3). The new infection rate has precipitated from 30 percent annually in 1999 to 44 percent in 2003 (Kaufman, Kleinman, and
Saich 2006b, 3; B. Wu 2004). UNAIDS warns that the disease, if left
unchecked, could afflict 20 million Chinese by 2010 (Gill, Chang, and
Palmer 2002, 97). Clearly, China is confronting an epidemic of daunting
proportions.

Since the early 1980s, the booming commercial sex industry and
the concomitant epidemic of sexually transmitted diseases (STDs) have
fueled and intensified the HIV epidemic. Following the discovery of the
first HIV case in 1984, between 1985 and 1988, a small number of foreigners and overseas Chinese in coastal cities were found to be infected.
From 1989 to 1993, intravenous drug users (IDUs) were found to be
infected by HIV in Yunnan province (Hyde 2007; Kaufman and Jing
2002). Beginning in 1994, HIV spread from Yunnan to thirty provinces
and regions through blood transfusion and sexual transmission. In 2000,
an HIV epidemic was reported among commercial plasma donors in
several provinces. In Henan, poor peasants supplemented their meager
income by selling blood at local stations. After their blood plasma was
removed in poorly sanitized centrifuges, the remaining blood was rein-
jected back into their veins. In one village, Wenlou in Shangcai County,
more than 60 percent of the population became infected with AIDS in
this way. Between 1996 and 2000, the reported number of sexually transmitted diseases in China doubled, reaching eight hundred sixty thousand
reported cases in 2000 (Jiang et al. 2002; Kaufman and Jing 2002; Parish
et al. 2003; Zheng et al. 2000).

Beyond the two groups of IDUs and blood donors, a sexual epidemic
has begun (Kaufman, Kleinman, and Saich 2006b, 3–4). As reported, a
major route of new HIV transmission is drug injection and commercial
sex, especially heterosexual contact (Longde Wang 2007, S5). While in
2004, sexual transmission accounted for only 31 percent of all reported
infections, with 20 percent attributed to heterosexual and 11 percent to
homosexual transmission, the epidemic that began with drug users is now
increasingly spread through sexual contact (Kaufman and Meyers 2006,
50; Qu et al. 2002; Van den Hoek et al. 2001; Wang et al. 2001b; Xiwen
et al. 2000; Zheng et al. 2000).[1] This phenomenon has been identified
as a product of widespread prostitution, changing sexual behaviors and
norms, internal migration, and low knowledge about transmission routes
(Kaufman, Kleinman, and Saich 2006a, 3–4; Zhang et al. 2002).

Until 2002, the government's overall response to AIDS was silence
and denial (Gill, Chang, and Palmer 2002). This response shifted to

commitment by the end of 2003, signaled by the China CARES program implemented by the State Council in March 2003, designed to provide care and treatment to those most affected by AIDS (Kaufman, Kleinman, and Saich 2006b). In February 2004, the State Council formed a multi-sector State Council AIDS Working Committee chaired by Vice Premier Wu Yi, soliciting participation of senior officials from all provinces to address the AIDS issue. This new committee implemented a policy of "Four Frees and One Care"—free antiretroviral treatment, free counseling and testing services, free treatment for pregnant women and testing for their babies, free school fees for orphans, and financial support for affected families (Kaufman, Kleinman, and Saich 2006b, 4). Since then, AIDS has become a priority of the State Council (Yip 2006, 179–80).

In the society, more than two hundred nongovernmental organizations (NGOs), both local and international, are at work to deal with HIV/AIDS in China. These organizations include Action Project against AIDS, launched by Wan Yanhai—the most prominent AIDS activist in China,[2] GONGO—Government-organized nongovernment organizations (puppet NGOs actually run by government officials), research centers, networks and forums, and clubs and salons with restricted memberships (Micollier 2005b).

Despite government efforts to disseminate information and reduce the stigma associated with AIDS through the policy of "Four Frees and One Care," China faces pressing challenges in implementing the policy, raising HIV awareness, and reaching and treating sex workers and their clients, as well as drug users (UNAIDS 2006). Research has demonstrated deficiencies in the surveillance and care reporting system, paucity of social science evidence, lack of coordination between Centers for Disease Control and Prevention (CDC) and hospital systems, poor health-worker attitudes, overcharge of drugs and services by health providers, deteriorating health education and preventive health care, lack of health insurance coverage for most rural citizens, and other limitations that have thwarted the government's endeavor (Kaufman and Meyers 2006).[3]

One predicament that these interventions must cope with is the low level of condom use. Consistent condom use has proven to be the most reliable method to prevent the spread of HIV/AIDS and STDs. Yet a 2003 national survey revealed that 17 percent of Chinese had never heard of HIV/AIDS and 77 percent did not know that condom use could prevent transmission (Hunter 2005, 91). Even those who are informed may not engage in safe sex. Research has shown that although men who engage in commercial sex are usually more educated and affluent (Parish and Pan

2006; Zhang 2001), condom use in the sex industry is consistently low (Parish and Pan 2006; Rou et al. 2007, 96; Wang et al. 2001a; Wu et al. 2007, 90). The estimations are that 56.7 percent of men who patronize female sex workers consistently use condoms in high-class hotels, while only 29.9 percent use condoms in bathhouses, massage parlors, and barbershops. For streetwalkers, the figure is an appalling 15 to 20 percent (Parish and Pan 2006). In booming manufacturing centers such as Shenzhen, the 25 percent annual increase of HIV infection is largely due to unprotected sex (Kaufman and Meyers 2006, 51). As Kaufman, Kleinman, and Saich note, "Unaddressed risk factors for sexual transmission of HIV are already present and will drive the epidemic in the coming years" (2006b, 9). As "an effective and targeted response to the epidemic will require higher-quality data and a more sophisticated policy research function" (Kaufman and Meyers 2006, 48), this book will help fill this void and address the underpinning risk factors that are inherent in commercial sex in China's sex industry today.

HIV/AIDS IN DALIAN

Compared with the national AIDS situation, Dalian has two distinctive local characteristics. First, the major transmission route is sexual conduct. Second, there is a rapidly increasing rate of infected women. The statistics showed that the rate of infected women in Dalian was more than 40 percent of total infections, more than twice the national rate of 18 percent. Moreover, HIV infections are routinely discovered from hospital visits (see also Y. Jiang 2005).

According to my interviews with some staff members in the local CDC, in Dalian, HIV was first identified among returnees from abroad in 1992, followed by two male migrants who attempted to sell their blood in 1995 (see also Anon. 2006c; Tong and Li 1996; Tong and Yang 1998). Since then the number of HIV infections has been growing. According to the staff, 2005 witnessed a fast growth of HIV infection, and the following year HIV infections grew even faster. In 2005, the official statistics for HIV infection in Dalian was more than one hundred. In 2006, the infection rate increased by 46.6 percent from the previous year.

In July 2006, Dalian was designated as the AIDS project city, sponsored by the Chinese Fifth Global Fund (Anon. 2006b). As a result of the project, Dalian has established forty labs for AIDS research and twelve volunteer test centers. According to the Dalian AIDS prevention administration, in consonance with the AIDS prevention stipulations promulgated in 2006, billboards were erected in the city to "transmit AIDS knowledge to high-risk populations." Automatic condom machines with twenty-four-hour service were installed in public places, such as seaports, train stations, bus stations, streets, and residential places (see also X. Zhao 2006). Moreover, following the nationwide surveillance system established in 1986,[4] 106 sentinel surveillance sites had been established in Dalian by the end of 2007 and 4.26 million people had been tested (see also Y. Shen 2007). In addition to mandatory testing, the local CDC also implemented an anonymous mail-order volunteer test, whereby patients can collect their own blood from fingers or ears and mail the blood sample to the AIDS lab (Sun 2007).

Studies have shown that the STD rate in Dalian was the highest and the second highest in the Liaoning province in 1997 and 1998, respectively (X. Li 2004; Yang and Li 1999; Yang and Zhang 1997). Concomitant with this high sexually transmitted infection rate, the major route of HIV transmission in Dalian is sexual contact, accounting for 74.19 percent of the total infections (Sun 2007). Male patients outnumber female patients. The majority of the HIV-infected population is the urban male customers involved in the entertainment sex industry (see also X. Zhao 2006). The HIV-infected population ranges from nine-year-old children to forty-five-year-old adults, with 88 percent of the infected below age forty-five (X. Zhao 2006).

In 2001, the State Council issued Law No. 40, "Plans for Prevention of AIDS in China (2001–2005)," which proposed 50 percent condom use in "high-risk groups" by the end of 2005. The State Council subsequently promulgated laws No. 7 and No. 248 in 2004 and No. 457 in 2006. According to the No. 457 law addressing AIDS prevention, entertainment places that fail to provide condoms or to set up condom sale establishments will be fined, closed, or have their business permits confiscated. These laws led to the initiation of the program of 100 percent condom use in entertainment places in provinces such as Hunan and Gansu and in cities such as Chongqing, Changsha, Xi'an, Zhengzhou, Hainan, and Beijing. The Dalian government issued Law No. 2 in 2005, mandating 85 percent condom use by 2005 and 100 percent condom use by 2006 in entertainment places.

BIOSOCIAL AND POLITICAL STUDY OF HIV/AIDS

In addressing the AIDS epidemic, the current dominant biomedical model explains health as stemming from individually chosen lifestyles.

Epidemiologists focus on individual sexual behavior and apply value-neutral objectivity and rely heavily on surveys as their primary method of scientific study (Bastos 1999; Frankenberg 1994; Hunt 1996; J. Mann 1996; Schoepf 2001, 339). Following the rational choice model, public health studies tend to assume that people engage in dangerous behavior because they fail either to recognize or to underestimate the risk involved in such behavior. It is also assumed that if individuals are informed of these risks, they will recalculate and abstain from such behavior.

Studies on sexual behaviors have focused on surveys of risk-related sexual behavior. They have aimed to collect quantifiable data on numbers of sexual partners, the frequency of different sexual practices, previous experiences with other STDs, and other similar issues related to HIV infection. The underlying assumption was that individual behaviors could change because of knowledge gain and the application of logical reason. However, studies throughout the world have discovered that increased knowledge of AIDS did not translate into widespread protection and that interventions based on information and reasoned persuasion had apparent limitations (Carrier and Magana 1991; Clatts 1989; Herdt and Lindenbaum 1992a; Herdt et al. 1991; Schoepf 2001, 343). This dominant biomedical approach dismisses the fact that sexual behaviors are related to social conditions and shaped by cultural systems and therefore is "unable to deal concretely with the lived social realities" (J. Mann 1996, 3; Parker 2001).

Because different societies and different subcultures within the same society exhibit different understandings of sexual expression, it became apparent that a far more complex set of social, structural, and cultural factors that mediate the structural risk in every group need to be taken into account to fully explain sexual conduct (Adams et al. 2005; Clatts 1994; Fausto-Sterling 2000; Herdt 1996; Herdt and Lindenbaum 1992; Parker 1994; Singer et al. 1992). In this book, I eschew the mechanistic assumptions inherent in the "rational choice" model widely employed in public health and adopt a more complex, socially inflected view of human behavior (see also Bolton and Singer 1992; Carrier 1989; Flowers 1988; Herdt and Boxer 1991; Obbo 1995; Parker 1987, 1988).

This book integrates cultural and structural concerns to provide an alternative to the previous overarching paradigm. Since diseases and epidemics are social processes shaped by political economy, social relations, and cultural factors, this book adopts a biosocial and political approach—a "political economy-and-culture" strategy, to emphasize the sociocultural and political dimensions of disease transmission and prevention (Schoepf 1998; Singer 1998, 1994). It aims to contribute not only to the development of HIV/AIDS research and intervention policies but also to anthropological inquiries on sexuality, gender, HIV/AIDS, and the power structure.

CULTURAL MEANINGS OF SEXUAL BEHAVIORS AND HIV/AIDS

There has been a growing focus in anthropological works on the interpretation of cultural meanings and the impact of structural factors as opposed to calculation of behavioral frequencies, as central and crucial to fully understanding sexual transmission of HIV (see Micollier 2004a; Parker 2001).

Following this analytical line of inquiry, I explore a range of cultural factors, including the cultural meanings related to male sex consumers' sexual behaviors as critical to an adequate understanding of the social dimension of HIV/AIDS. My study of male sex consumers shows that most are relatively well informed about the risks of unprotected sexual contact; yet, many still refuse to use condoms. My research strongly suggests that cultural factors, such as concepts of masculinity and male-female relationships, exert a strong influence on male consumers' decision making about condom use. Other cultural factors such as the clients' perception of contraceptive use as the woman's responsibility, and the perception that rejection of condoms exhibits "bravado" or "valor" as perceived by the peer group all come into play.

Many men regarded rejection of condoms as a political act of defiance. During my research, male clients repeatedly explained that they insisted on achieving absolute sexual pleasure without condoms. They resented the state's attempts to regulate their pursuit of sexual pleasure and expressed hostility at mention of HIV/AIDS and condoms. To them, these are the state's tools and weapons to control and police their sexuality. AIDS and condoms represent exactly what they rebel against—control and regulation of their sexual pleasure. Warnings about AIDS and condom use shackle their sexual pursuits and thwart their absolute enjoyment of sex, free of any moral constraints. They reject mainstream morality and celebrate sexual freedom by rejecting the state's constraints.

My focus on sexual meanings emphasizes their shared and collective character. I examine and explicate what sexual practices mean to male clients, the cultural contexts in which they take place, the social script of sexual encounters, and their sexual culture. Beyond calculation of behavioral frequencies, my work recognizes that people are not atomized or isolated individuals but, rather, social persons who are integrated in a specific cultural context (see also Herdt and Lindenbaum 1992b). By emphasizing their own cultural concepts that structure and define their sexual experience and sexual practice, my cultural analysis moves from an etic, or outsider, perspective to an emic, or inside, perspective, "from the experience-distant concepts of biomedical science to the experience-near concepts and categories that the members of specific cultures use to understand and interpret their everyday lives" (Parker 2001, 167). My ethnographically grounded descriptive and analytical research on the social and cultural construction of sexual meanings will shed light on understanding of sexual risks, and facilitate the development of culturally sensitive and culturally appropriate, community-based prevention programs (Parker 2001, 168).

Another factor involves the cultural meaning of AIDS. Internationally, AIDS is "an epidemic of signification" (Triechler 1987). Moral and stigmatizing responses proliferate throughout societies. Following Foucault, such dissipated knowledge from the powerful is socially and politically situated, built on knowledge of the power to define what and determine how the public know (Schoepf 2001, 338). For instance, studies in Africa have shown a sharp contradiction between the severity of the epidemic and the discourse that supports and reproduces gender, class, color, and national hierarchies (Schoepf 1998). Early responses constructed AIDS as a disease invented by Westerners to discourage Africans from sex and procreation (Schoepf 1998, 341; Smith 1996).

Like Africa, societal responses to AIDS in China are propelled by cultural politics, not only forged in the history of relations between China and the West but also couched in the construction of the category of "high-risk groups" along the lines of gender and sexuality. Also, as in Africa and other countries such as the Caribbean and the United States, discourse in China stigmatizes prostitutes and immoral women as the diseased and the dirty, a reservoir of infection, from whom other "pure" people need protection (Brandt 1988; Lyttleton 1996; Taylor 1990a). "Creation of alterity allows those in power to demonize, to scapegoat, to blame, and thus to avoid responsibility for sufferers" (Douglas 1991; Schoepf 2001). Farmer calls this social stigma and discrimination "a geography of blame" (Farmer 1992).

A focus on risk groups rather than risky behaviors not only carries the wrong message to the public, making people outside of the categories believe they are not at risk, but also serves to demonize, blame, and stigmatize certain groups (see Lyttleton 1996; Parker 1987). In the history of HIV/AIDS, gay men, Haitians, and Africans were placed in the high-risk category, stripped of other identities, discriminated against, and dehumanized. Farmer notes that Haitians were denied housing, dismissed from jobs, and tested before their entry into the United States. The purpose of the construction of high-risk groups is to reinforce the "hegemonic process that helps dominant groups to maintain, reinforce, re-construct, and obscure the workings of the established social order" (Glick-Schiller 1992; Schoepf 2001, 338). To unravel this hegemonic process, in Chapter 2, I will delve into a critical analysis of the hegemonic discourse in China that reasserts male hegemony, promotes a xenophobic nationalism, and justifies social inequalities in the configuration of gender and sexuality.

As Paul Farmer notes, it is "urgent to have a proper biosocial understanding of AIDS in China" because of "a shortage of social, economic, and behavioral research in planning the China AIDS response" (Farmer 2006, xxi). As illustrated, my research departs from previous frameworks that emphasize the "problem populations" of sex workers, "rational-choice" that attributes high-risk behavior to inadequate knowledge, and questionnaires that disregard personal contact. Instead, this book fills in the lacuna called for by Farmer and employs a biosocial study of AIDS that emphasizes the cultural meanings of AIDS and the sociocultural context of male clients' sexual behavior and male-female gender relations.

POLITICAL AND ECONOMIC ANALYSIS OF STRUCTURAL FACTORS

Anthropological literature on AIDS from the 1990s has revealed not only local sociocultural processes that create risks of infection but also the political economy that generates structural violence of poverty, powerlessness, and other inequalities of class, gender, and ethnicity (Schoepf 2001). Research has shown that not just cultural but also structural, political, and economic factors have played a key role in shaping sexual behaviors and determining the spread of the epidemic (Baer, Singer, and Susser 1997; Farmer 1999; Farmer, Connors, and Simmons 1996; Farmer et al. 1993; Lindenbaum 1997, 1998; Schoepf 1991, 1995; Singer 1998; Singer et al. 1990; Singer et al. 1992; Singer 1994). The structural, political, and economic factors have been described as forms of "structural violence," determining the social vulnerability of groups and individuals as a result of poverty, economic exploitation, gender power, sexual oppression, racism, social exclusion, and so on (Farmer, Connors, and Simmons 1996; Parker et al. 2000, 2001; Singer 1998). As Parker notes, risks of AIDS "can never be fully understood without examining the importance of issues such as 'class,' 'race' or 'ethnicity' and the other multiple forms through which different societies organize systems of social inequality and structure the possibilities for social interaction along or across lines of social difference" (Parker et al. 2001, 169).

Studies of Africa have demonstrated how local economic contexts have shaped the status of women, and youth, and gender relations (Setel 1999). "AIDS has struck with particular severity in communities struggling under the burdens of poverty, inequality, economic crisis, and war." Many people who are well aware of the danger of sexual transmission, cannot avoid becoming infected because they cannot control the relations of power that put their lives at risk (Schoepf 2001, 336). AIDS came to stand for "acquired income deficiency syndrome," "a disease brought on by poverty, unemployment, and the strategies that poor people commonly adopted for survival" (Schoepf 2001, 341).

Structural violence can be identified as the force that compels the poor and disempowered to choose survival strategies that have contributed to sexual risks (de Zalduondo and Bernard 1995; Farmer, Connors, and Simmons 1996; Hammar 1996; Kammerer et al. 1995; Schoepf et al. 1988; Schoepf et al. 2000; Setel 1999). For instance, Farmer shows that the construction of a hydroelectric dam funded by USAID drove peasants from their lands, causing poverty and increased risk of AIDS (Farmer 1992). A similar process in Ghana drove women from landless families to Côte d'Ivoire, where the only work available was prostitution. Many contracted HIV (R. W. Porter 1994). Therefore, many social scientists have contended that transformation of the living conditions of the world's poor and disenfranchised is indispensable to curb the AIDS epidemic (Schoepf 2001).

In China, structural violence is situated in the historically constituted political and economic systems in which political and economic processes and policies related to economic development, housing, labor, migration, health education, and so on, create the dynamic of the epidemic. Farmer observes that China's AIDS epidemic "has its roots in the political and economic transformations that began in the late 1970s" (Farmer 2006, x). More specifically, rapid change has spawned uneven rural versus urban economic growth, and polarization of urban and rural populations in the realm of wealth and health. This, in turn, has induced different trajectories in the urban and rural epidemic of AIDS—the former through sexual transmission and the latter through blood transmission.

Research has shown that China's epidemic is closely associated with social and economic disadvantage, especially in the configuration of gender, rural-urban residence, and poverty (Hyde 2007; Kaufman and Meyers 2006; Saich 2006; Shao 2006). For instance, Sandra Hyde (Hyde 2007; Kaufman and Meyers 2006a; Saich 2006; Shao 2006) argues that while Chinese public health officials and social scientists respond to the epidemic by linking HIV/AIDS to ethnic cultures and their cultural behaviors, in reality, it is poverty and drug trafficking that drives the epidemic in this area. Hyde contends that the Chinese government is unwilling to acknowledge the problem of poverty and, instead, blames it on the minority borderlands because Han Chinese define ethnic groups as loose and sexually uninhibited, and their practices as illegal, unsafe, and unhealthy. The underlying goal is to control China's borders through controlling borderland bodies.

In rural China, a review of twenty-year trends in underinvestment in health in rural China demonstrates that political policies during economic transformation have discriminated against the poor and taken a toll on the urban unemployed and rural migrants in the city (Saich 2006). Vast income inequality and unequal distribution of resources in the rural versus urban areas have led to poor-quality public medical care and much fewer numbers of health personnel and facilities in the rural areas. It is no surprise that according to the government's own statistics 80 percent of AIDS patients are rural residents (Yip 2006, 186).

Despite an AIDS prevention policy that looks good on paper, a plethora of challenges obstruct its effective implementation (Liu and Kaufman 2006; Saich 2006). For instance, the government circumscribes media health education; mistrusts, resists, and exerts control over NGOs; and monopolizes the society with government organized NGOs (GONGOs)—puppet NGOs actually run by government officials (Micollier 2005b; Ru 2006; Saich 2006). The government's spending on public health also dropped over the years, leading to the limited capacity of the health system to respond to AIDS, shortage of doctors and trained personnel, expensive HIV tests, ineffective implementation of free treatment, and insufficient funding. The Chinese government's investment in HIV prevention only amounts to one-seventh of that of Thailand (Dechamp and Couzin 2006; Liu and Kaufman 2006).[5] National policy is also subverted when local governments cover up the spread of HIV because of their weak incentive to engage in AIDS prevention and their concern about the effect of openness on HIV (Saich 2006).[6]

In this book, I discuss some of these structural factors that hinder and thwart AIDS prevention. In Chapter 3, I unravel the controversy and sensitivity about the advertisement and distribution of condoms and the insistence on abstinence-dominated education and confinement of education to mere anatomic and physiological knowledge (see also Ru 2006). Fear that sex education and marketing of condoms would disrupt and collapse the country's value system has led to contradictions between policies and implementations (see Ru 2006). I also examine what Micollier couches as "the ongoing debate about sexual control and object of official concern," that is, the conflict between state management and control of sexuality and the sexual revolution and health concerns under the influence of globalization and the consumer revolution (see also Micollier 2005a, 2006; Rofel 2007). Regarding the dynamics between male sex consumers and female sex workers, I will delve into the underlying power hierarchy between urban men and rural migrant women, historically antirural political policies such as the household registration system that have culturally, socially, and politically segregated and discriminated against rural people and created a rural-urban apartheid. Once the urban gate was open, 1.2 million rural people flooded into the city seeking a better living. Unfortunately, many of the women migrants were only able to survive as sex workers, leaving them vulnerable to HIV infection (see T. Zheng 2003, 2009).

GENDER AND SEXUALITY

Gender power differentials have been identified as central to a better understanding of the structural factor, as research has shown that the political economic factors driving the epidemic are intertwined with gender hierarchy and sexuality, wherein women, and low-income women in particular, are vulnerable to HIV infection (de Zalduondo and Bernard 1995; Farmer 1992; Gupta and Weiss 1993; Micollier 2004a; Parker 1991; Schoepf 1992; Sobo 1993, 1995a, 1995b, 1998). Studies of impoverished inner-city ethnic groups and the urban poor in the United States have demonstrated that gender-power inequalities interact with other forms of structural violence, such as poverty and racism, thereby creating extreme vulnerability for the disenfranchised (Farmer 1996; Singer 1994, 1998; Sobo 1993, 1994, 1995a).

Farmer has noted that sexuality is a poorly understood topic by nearly all social scientists and that AIDS intervention programs often rely on "superficial rapid ethnographic assessment procedures rather than on more detailed ethnographic description and analysis" (Farmer cited in Parker 2001, 170). This study fills in the gap and provides an ethnographic understanding of the political economy of HIV/AIDS in the urban Chinese sex industry. In particular, it delves into the intersection between poverty, gender-power hierarchy, and political policy that structures sexual practices and condom use between male sex consumers and female sex workers. More specifically, this study illustrates how sexual inequality between male sex consumers and female sex workers determines condom use or nonuse; how historical, social, and cultural rules and obligations shape patterns of contraceptive use; negotiations of condom use in sexual encounters; and possibilities of sexual violence.

In China, where women are culturally expected to engage in sex only within the bounds of marriage, imbalances in gender power and social constructions of gender prevent women from negotiating safe sex even within marriage (cf. MacPhail and Campbell 2001; Holland et al. 1992a, 1992b). Sexual double standards that sanction male infidelity and attribute condom requests to female infidelity or disease fuel men's risky behavior that infects women (see P.-F. Kelly 2004; Micollier 2004b; O'Leary 2000; Walters 2004).

Moreover, as I will demonstrate in Chapter 1, the nature of family planning and priorities in contemporary China continues to place women at risk. Throughout Chinese history, female infanticide was common during periods of famine. Today, as a result of conflicting pressures to produce a male heir and to adhere to the guidelines of the state's one-child policy, female infanticide and abortion of female fetuses have once again become common (Greenhalgh and Winkler 2005). Although a shift of policy in 2000 purportedly favored women's reproductive health and female control over contraceptive choices, entrenched cultural ideas of gender inequality and consistently misogynist policies continue to shape actual practices.

My research helps clarify the understanding of gender relations and power structures revealed with condoms and disease control. As is common throughout the world, even in societies in which relatively egalitarian sexual relations are the norm, men in China typically decide whether to use condoms (Evans 1997; Farrer 2002). This male power is amplified in the case of sex-money exchanges in which the female sex worker depends on the man for her livelihood. My research reveals that, in many cases, women sex workers, even those who were informed about the dangers of HIV/STDs and the risk-reduction benefits of condom use, were unable to convince their clients to use condoms. Ironically, of course, nonuse of condoms exposes not only women hostesses but also their male clients to potentially life-threatening diseases. When the condom is not an option to protect against infection or pregnancy, hostesses turn to other methods such as presex shots, emergency contraceptives, liquid condoms, external-use liquids, and so on. However, these do not provide protection against STDs and HIV. The female condom has not been widely used in China because of its high cost and unavailability. In fact, only two domestic factories produce female condoms in China (P. Zhao 2005).

This study merges cultural and political economic approaches against the dominant biomedical perspectives. I define structural or political economic forces as political policies, economic differences, gender hierarchies, and power relations. This study demonstrates how cultural orders and social policies in China impose different rules and obligations on men and women and how differential economic and political power relations underpin sexual practices and condom use between male sex consumers and female sex workers. This study casts the bodies of female sex workers and male sex consumers as "a symbolic and material product of social relations—a construct that is necessarily conditioned by a whole range of structural forces" (Parker 2001, 171).

RESEARCH METHODS AND LIMITATIONS

This ethnographically contextualized study provides a unique perspective on the study of disease transmission grounded in the everyday lives and worldviews of clients and hostesses. During my two-year fieldwork between 1999 and 2002, I lived and worked with karaoke bar hostesses as a karaoke bar hostess myself. In my previous book *Red Lights: The Lives of Sex Workers in Urban China*, I discussed the trials and tribulations throughout my fieldwork (Rofel 2007). By working and living with these bar hostesses, I learned what it was like for these young women to struggle for respect, to improve their social standing, and to fashion themselves as modern women. During this period of my fieldwork, I was deeply disturbed by the fact that many male clients abjured the use of condoms. As a result, many hostesses suffered from frequent abortions and STDs.

I went back to the field in 2004 and conducted fieldwork off and on for twelve months until 2007, on condom use, sex work, and HIV/ AIDS, among the same population of hostesses and male clients. During this time, I went back to the three karaoke bars I had worked at during my previous fieldwork from 1999 to 2002 and resumed contacts with the medium-tier bar owner and bar manager, more than fifty clients and more than fifty hostesses.

Although my previous research was based on more than 200 hostesses, upon my return in 2004, some were married with kids and had returned to their rural hometowns; some had left Dalian for Shanghai, Singapore, or Japan for more profitable hostessing work; some had changed their workplaces and worked at different karaoke bars; and others had shifted their work to engage in massage and sex work at local sauna bars. Some of my closest informants worked at sauna bars upon our reunion. They told me that various reasons led to their change of work. First, their health, especially their liver and kidney, deteriorated so much because of overdoses of alcohol at karaoke bars that their bodies simply could not hold it any further. Second, hostesses told me that male patrons were fewer than before and, as a consequence, they earned much less than before. They said, "Now we have to sit [zuotai—accompany customers] all night with one customer, sometimes as much as 8 hours—from 7 p.m. until early in the morning. Previously we sat [zuotai] six or seven times a night, earning six or seven times more than today." When I learned how much they earned every month, at least 3,500 to 4,000 yuan a month, I commented that that was pretty good. They rejected my response immediately and said, "No, not good at all. We came out to earn money, not to make ends meet. No one wants to just get by in life. We want to live a better life and earn lots of money." The slackening business at karaoke bars was another reason that hostesses left for sauna bars, where sexual service was more direct and less lengthy. Another factor was the influx of newcomers who were so young that they were born after the 1980s. Hostesses felt lost because they were too old to compete with them. Hostessing is a young woman's profession. One must make a lot of money in a short time if one is going to make a transition into a more suitable urban lifestyle. Some of my close hostess informants were forced to leave karaoke bars and were hired by sauna bars as masseuses.

Through a handful of my close hostess and client informants from my previous fieldwork, I managed to meet and reconnect with over fifty hostesses and fifty clients. This book allows the voices of at-risk people to be heard, by incorporating active participant observation, texts of interviews, personal stories, surveys, and questionnaires.

PARTICIPANT OBSERVATION

I engaged in long-term active participant observation of interactions between clients and hostesses, everyday lives of hostesses, and socializing activities of clients. For instance, I followed clients and attended their group activities such as playing golf, singing songs at karaoke bars, consuming at sauna bars and observing interactive behaviors of male consumers and female hostesses, dining at restaurants, chatting at coffee shops and teahouses, and so on. I also stayed with two hostesses at their apartments for a while, shopping with them, eating out, singing songs at karaoke bars with them and their clients, watching TV, going to hairdressers, accompanying their friends at apartment rental services, and other everyday activities. I also took care of a hostess friend for about three weeks after her abortion. Daily, casual interactions with hostesses and clients not only helped me reestablish rapport and trust with them but also became an important source of data.

Aside from the fieldwork with the clients and hostesses, I also spent a couple of weeks serving as a sales clerk at a local condom company, trying to understand the condom market in the local area. As a sales clerk, I fulfilled my responsibilities every day by jumping on city buses, traveling from pharmacies to convenience stores around the city. At each store, I interviewed the salespersons at the counter about the sales volume and the most desirable condoms and checking the condom display location on the sales counter. The condom company manager informed me that every condom company provided bribes to the managers and salespersons of the stores, so that the salesperson could push their brand to their customers and exhibit their condoms at the most highlighted place on the sales counter—the center. If their condom was not placed at the center, I had to report it to the company manager because it meant that other companies had offered more bribes, and we needed to increase our bribe.

In addition to this segment of fieldwork, I committed a month and a half to active participant observation as a counselor at a local AIDS NGO to understand the local cultural milieu of AIDS organizations and their interrelationships and the attitude toward AIDS education and AIDS patients on the ground, both in grassroots organizations and in government bureaus. At the local NGO, most members were gay men, and a couple of NGO volunteers were HIV patients. As a counselor at the NGO, I was responsible for negotiating with the local CDC and the Red Cross for AIDS education funding. I attended their intense study meetings on drug abuse and HIV/AIDS, participated in an AIDS walk around

the city, talked to them and learned their personal stories and sexual practices, and observed them surfing on the gay Web sites, chatting, and pick-
ing up sexual partners.

SURVEYS

I designed a survey on HIV knowledge, condom use, and attitudes toward
AIDS patients for eight groups: hostesses, clients, female and male service
workers, female and male college students, and female and male professors. My intention was to compare and contrast the answers and find
out which group had the highest HIV knowledge. One of my previous
contacts, an owner of a well-known four-star hotel in the city, helped me
distribute one hundred surveys to the male and female service workers in
his hotel.

I also solicited help from the president of the local medical university and a professor friend at the Maritime University to obtain survey
answers from their professors and students. They turned in survey results
from one hundred professors and one hundred students from both universities. At the Maritime University, I was told that surveys would not be
distributed to the students unless I deleted words such as "oral sex" and
"anal sex." So I did. The surveys came back with these "sensitive" phases
crossed out with black markers and certainly without answers. The professors' "protective" attitude toward students shocked me at first, but I
was not surprised later when I taught at the Capital Medical Science University in Beijing and realized that those medical students were ignorant
of whether oral sex could transmit HIV.

In Beijing University, I attended the first youth AIDS education film
with the university students, which I will discuss in detail in Chapter
2. Before the film, students around me held the distributed booklet of
the synopsis of the movie, asking one another what HIV-positive meant.
They were debating whether HIV-positive or HIV-negative meant the
person was sick. Some said HIV-positive meant sick, whereas others said
HIV-negative meant sick. They were frustrated that they could not reach
a consensus.

As time went on, because of the pervasive ignorance of HIV/AIDS that
I encountered every day during my research, I became used to it. I was
no longer surprised when my two hostess roommates were certain that
Pu Cunxin, the image ambassador of AIDS intervention in Dalian and
in China, was an AIDS patient. The rationale was that no one would be
involved in the HIV/AIDS cause unless he were an AIDS patients. They

assumed that only if Pu Cunxin was an AIDS patient would he have a stake in the AIDS intervention business. Therefore, he was outspoken on this issue. Other hostesses believed that there was a cure for this disease.

To systematically compare and contrast HIV knowledge among these various groups, I needed to elicit survey results from both hostesses and clients. As I resumed contact with the male consumers and hostesses from my previous research, I talked to them about my new research and asked for their help with the surveys. I gave out ten to twenty surveys to each of them so that they could circulate them to their friends to fill out. It took them a couple of months before they returned the surveys. At first, I would meet with them and ask whether the surveys had been disseminated to their circle of friends. They told me that they needed time, as it was not an easy task. They commented that the questions on the survey were "too private and too sensitive," so they felt embarrassed to hand them out to their friends. One hostess jokingly said, "What a hell of a task you have given me! It's going to take me a while before I can get twenty other hostesses to fill them out!" I was deeply grateful for their earnest and sincere help with my research. In the end, they were able to return one hundred completed surveys.

This process was not smooth. One male client resisted filling out the survey himself, let alone distributing it to his friends to fill out. He was the owner of an import-export company. When I approached him for help, he angrily refused and said, "Chinese people think differently from the Americans. I am telling you—they think differently. If I give them these surveys, they are going to think that I am an HIV/AIDS patient. I am telling you—I can't do this. People these days are extremely repulsed by this kind of thing. I am sorry, but you can't force me."

Refusals to answer questions and expressions of anger and embarrassment from my research subjects permeated the my research. Coupled with similar issues arising from my interviews with male clients, these adverse reactions, as I will explicate in the next section on interviews, crystallized the nature of the cultural milieu surrounding AIDS in China.

Nonetheless, I received one hundred survey answers from cooperative male clients and hostesses. Yet, I was still halfway from my goal of two hundred survey answers. To accomplish that goal, I visited the local CDC to ask for possible collaboration to distribute one hundred questionnaires to karaoke bar hostesses and clients. The CDC director declined because they had no power to enter karaoke bars to access hostesses and clients.

After several futile attempts, I finally decided to seek help from the bar owner of the medium-tier bar I reconnected with from my previous research. Before I approached him, I asked myself: Will he allow me to distribute these surveys to hostesses and clients in his karaoke bar? What would this mean to him and to his bar? My prior setbacks and rejections during the process had prepared me psychologically to deal with further such occurrences.

I started frequenting the medium-tier bar to try to meet up with the bar owner. The bar manager told me that the bar owner was busy overseeing the construction of another business on the other side of the city and would not show up at the bar for a while. So I talked to the bar manager, who was also one of my informants during my past research, about my new project and asked whether he was willing to distribute the questionnaires to one hundred hostesses and clients in the bar. He looked at the survey questions and shook his head, saying, "I am so sorry that I am afraid that I cannot help. If I were to do this, it would frighten away our clients, and it would also put us in peril because the police could use it as evidence to claim that we engage in sexual transactions in the bar and close us down." The manager believed that distribution of surveys would lead clients to believe that hostesses in his bars carry STDs and HIV and, therefore, would stop patronizing. He also recounted that the police had forced a local karaoke bar to close down because condoms were marketed in the bar.

Although I was not surprised by his refusal to help out, I was disappointed and even more worried about whether I could fulfill my research goal, given such a tenuous and sensitive cultural arena. I finally decided to bother the bar owner despite his full-time commitment to his next business project. I started calling him to find out when I could talk with him. First, he professed that he was too busy to add another commitment. I persisted and continued to call him for appointments. Finally, late one night, I received his phone call, telling me to meet him at the construction site on the other end of the city. Although it was already close to midnight, I took a taxi to the construction site. Both the bar owner and the bar manager were there. The bar owner told me that they had a respite from then until the next morning, so we could talk. I told him that I would prefer speaking with him individually. He asked the bar manager to leave the construction site. The construction site was a mess, with no proper seats or amenities. We sat on a stool to talk. I offered him a present I brought back from the United States. He thanked me and then came right to the point, "What do you want from me?" I explained the significance of the HIV/AIDS issue and condom use in society. I was nervous and unsettled, constantly fearing that he would interrupt me and give me the same disappointing answer. Surprisingly, he never interrupted me. In fact, he nodded and carefully listened to my speech. I don't know whether he was moved by the fact that I traveled an hour at night to meet with him, was touched by my indignant lecture, or maybe both. Whatever it was, at the end of my speech when I asked for his help to dispense the surveys, he looked through the questions and gave me an unexpected answer, "This is a noble cause. We should support it. However, we have to do it secretly so that the police would not know." I immediately agreed and thanked him profusely for his help. He summoned the bar manager over and asked him to see to it that the surveys were answered carefully by all the hostesses and some male customers. The next day, I brought more than one hundred surveys and one hundred little gifts for each respondent. I was in the bar when the madam handed out the surveys to the hostesses. The madam reminded the hostesses to answer carefully. After several days, I retrieved another one hundred surveys from the bar manager.

INTERVIEWS

I conducted open-ended interviews with local residents in general, HIV patients, clients' wives, and HIV doctors in a Beijing hospital specializing in HIV/AIDS, STD doctors in several Dalian hospitals, managers of two local condom companies, government officials from family planning offices,[7] Red Cross, and the local CDC, and sales personnel and owners of local adult health shops, convenience stores, and pharmacies.

The first week during my research, I visited the local CDC and interviewed some staff members. Then I volunteered to hand out AIDS education booklets and free condoms. The CDC director asked, "Where are you going to disseminate the booklets and condoms?" I replied, "Karaoke bars." She immediately looked disconcerted and worried. She repeatedly warned me, "You have to know the bar manager to do this. Otherwise, it would be too dangerous. No one wants these booklets or condoms." Since that was the beginning of my fieldwork, I was naive and unaware of the cultural complexity revolving around HIV and condoms. I said, "Why would no one want these education materials and free condoms?" She looked directly into my eyes, apparently surprised at my ignorance but responded, "Nowadays, people don't have the consciousness to learn about AIDS. People are strongly resistant and antagonistic towards it [dichu qingxu]. It will infuriate people if you give them this little book. They are very averse to this kind of information." I told her that I knew the bar manager. Only after she was assured that I knew the bar manager did she hand me some booklets. She said, "I can't give you any condoms. You are going there on your own [without protection from the authorities]. If you hand out condoms to the mangers, it will cause trouble. They may call the police and you would be arrested. We have had several volunteers in the past arrested by police because they handed out free condoms at karaoke bars. So don't distribute condoms. When handing out the booklet, be very careful. If they don't want it, don't force it. Remember: never force it. It's totally up to them whether they want it or not. Never force anyone to take the booklet."

I stood there, dismayed by her repeated, concerned, and lengthy cautionary words and slowly came to terms with the serious nature of what I had previously thought of as a "trivial" business. Handing out free stuff to people—why is it such a big deal that can lead to arrest?

I had to admit that I was a bit scared of replicating the previous volunteers' fate—police arrest. I did not want to repeat history, so I chose not to distribute these materials to random customers in karaoke bars. Instead, I handed them out to my male client and hostess informants. Since I was sharing an apartment with two hostesses, I offered to read the booklet to them. As I was reading, one of them interrupted me and said, "I don't want to know about this stuff. I am better off not knowing about this because otherwise I can't continue living." I had to admit that I understood their anxieties and frustrations. After all, they were unable to exert control over condom use with their client lovers. Under this circumstance, informing them of the possible consequence of AIDS did nothing but infuse fears and worries into their lives that they had no control over. Since they had to continue living this way and there was no other alternative, it was axiomatic that they would rather not know because knowing or not knowing did not make any difference. In this case, ignorance was bliss.

When I told them that the country would provide drugs free of charge, they commented with a sneer, "If you get it through donating blood, of course, the country will provide drugs for you for free. If you get it through sleeping around [guihun], there is no way the country will provide free drugs for you."

It did not take me long to realize that hostesses' negative attitude toward HIV information and their mistrust and disbelief of the government were typical of all informants involved in my research. Male clients were especially difficult to reach with information on HIV and condom use, albeit for different reasons.

For my own safety and the privacy of male consumers, I invited them to coffee shops and restaurants for interviews. Individual tables at coffee shops were pretty secluded and quiet, as were the rented rooms at restaurants, which provided ideal places for interviews of sensitive topics.

My interview questions were intended to better contextualize and situate their survey answers. I started out with questions that tested their HIV knowledge and the source of their knowledge, followed by questions about the source of their sexual education, their opinions about their sex education, their first sexual experience and condom use, and their sexual history. I asked when they started using protection, whether they were consistent in their practice of protection, their definition of sex and their control over sexual desires, the setting and occasion of their first intercourse, and subsequent sexual exchanges and so on.[8]

Undoubtedly, these questions generated an immense embarrassment and hesitation in the interviewees. The interviewees, understandably, were red-faced and smoking the whole time to cover up their embarrassment. Among all these questions, clients were most resistant to the topic of HIV/AIDS and condom use. For instance, when I initiated the interview with a question on HIV/AIDS knowledge, one client just brushed off the topic by saying, "AIDS in our country is mostly transmitted through blood transfusion." Although I corrected him and pointed out that heterosexual transmission was the leading cause, he did not show any interest or belief in my words. When I continued with questions about his practice of condom use, he said, "I don't care whether I use it or not" and ended this topic.

Clients' resentment and antagonistic attitude toward HIV and condom use created a frustrating predicament for my research. Conversations on this topic were hard to continue without clients' cooperation. Luckily, a few of them were supportive of my research and willing to sacrifice their discomfort during the interviews. However, they reminded me of what a "huge favor" they were doing me. They constantly told me, "If I were not your friend, I would not reveal the private side of my life! I am telling you these things because we are friends and I want to help out with your research." Another client said, "If anyone else asks me these questions, I would say 'I don't know.' Not because I don't know the answers, but because I don't want to answer them." Fortunately, we had been friends for years from my previous research, and they were willing to answer these "difficult" questions. Yet, at the same time, they never failed to grill me for answers to the following questions as well. "Can't you study another topic that is more pleasant than HIV/AIDS?" "Why

HIV/AIDS?" "Why something that none of us wants to think about?" "Are you aware that people are going to think that you have AIDS?" I always carefully explained the reasons for my research, while expressing my deep gratitude and appreciation for their active collaboration.

Despite these informants' active participation, the ongoing resistance to this topic plagued me. The truth is talking about AIDS and condoms turned people off. The clients loved to tell me all about their hostess girl-friends, but questions about AIDS and condoms silenced them. What was the reason behind this resentment? Why were these men never tired of talking about patronizing hostesses but repulsed by the topic of con-doms? What does a condom mean to them?

As my research progressed, answers to these questions revealed them-selves. Some male consumers had never used condoms in their lives, and some had only used them once. No wonder they were not "cooperative" when I asked them when they used or did not use condoms and where they purchased condoms. The response was, "I've never used one. So I'm not familiar with the brand names or the condom products in general." Others who did use condoms had no idea about the price or the brands of condoms. They never asked the price or selected condoms because it was too embarrassing. They simply asked the counter staff to get them a pack of condoms and left immediately after payment. For instance, one client explained it to me this way, "I am afraid of buying condoms. When I do, I drop my head like a criminal who has just committed a crime, hastening to escape after paying for the condom."

For these clients, condom use triggered their passionate anger about society's control of sexuality to which they attributed their embarrass-ment. For instance, one client adamantly said, "Sex is considered as a sin in the society. Talking about sex, even with your wife, is forbidden, is taboo, and is embarrassing." None of the male consumers in my research had ever talked about sex with their wives. One male client said, "Con-fucian ideas are still prevalent in people's minds. I've never talked about sex with my wife. It's too embarrassing to talk about it. In China, no one likes to talk about it [sex]. Even with my male friends, no one talks about sex. It's still a taboo. People are still conservative about these questions. They'll be reluctant to answer." He was also infuriated that condom use was legislated in karaoke bars, "Hostessing is illegal in China. How can the government legislate condom use? It's impossible [to implement it]."

To the clients, condoms symbolize loss of freedom. Condoms were not something they could brag about, like patronizing hostesses. Rather, condoms represented a hindrance, a barrier, to what they believed was a natural and innate sexual desire. Talking about condom use meant talking about controlling their sexual expression. Talking about condom use meant talking about curbing their sexual liberation. As one of the clients said in the interview, "In our personal space, we want hedonism. We want complete liberation. Condoms represent something that ties us up and prevents us from doing this. That's why we are so averse to this topic."

In the following chapters, I will discuss male clients' perceptions of condoms and their sexual culture. For the purpose of this section, for my research methodology, I have combined participant observation with formal and open-ended, as well as informal, interviews and surveys. Matthew Gutmann (2007, 27) observes, "There are practical limits to studying sexuality and sexual relations." In this study, the limitations include selection bias and reliance on self-reporting of sexual behaviors.

As illustrated in this section, the hostesses and clients who participated in my research were my previous informants and their friends. Given the sensitivity of the topic, I relied heavily on these key informants with whom I had developed a close relationship over the past decade. Because of this particular selection process of my research subjects, I was not able to reach a vast number of hostesses and clients in the city. This recruitment bias circumvented the breadth of research subjects and, hence, could potentially affect research findings.

Like all other researchers on sexuality, we are confronted with the limits of self-reporting. Unlike other social activities, sexual activities are private and reclusive. It is impossible for a researcher to be present, to observe, or to participate in my informants' sexual activities. However, although researchers have no other options but to rely on informants' self-reporting, I have found ways to enhance the reliability of the data. In my research, since I have developed a close relationship with both the hostesses and their client lovers and friends for a decade, I was able to corroborate their stories and detect and account for inconsistencies or discrepancies in their words and actions. Daily interactions with the hostesses and clients on a casual basis throughout a whole year's intensive fieldwork, not counting the previous two years of intense interaction, exposed me to a wide array of information and knowledge, either from group gossiping or through their family members that aided the verification process.

PREVIEW OF THE BOOK

Chapter 1 of the book casts my research in the broader context of gender politics imbued in practices of birth control, disease control, and control of women's bodies. More specifically, it explores the cultural obstacles to condom use through a diachronic discussion of birth control and disease transmission. I argue that the use of prophylactics and other contraceptive means in contemporary China reflects the historical continuity of gender hierarchy. Despite claims during the Maoist and post-Mao era about the equality and empowerment of women, the reality is a continuing deep cultural bias that holds women responsible for birth control and the consequences of the failure of birth control under conditions in which the state's one-child policy places heightened burdens on women to produce a male heir. This historical account helps foreground the current situation of control of women's bodies and women's responsibility for birth control with particular relevance to prophylactic practices.

Chapter 2 explores the underlying intricacy of male dominance in the media construction of HIV/AIDS. This analysis helps frame the topic of disease control within the context of controlling women's bodies and blaming women as vectors and transmitters of disease. Since the emergence of AIDS in China, discourses have targeted foreigners and immoral Chinese women as vectors and transmitters of the disease. The boundary of the nation is secured through blaming foreign nations as the source of diseases. The HIV/AIDS discourse allows the state to propagate a strict sexual morality in adult women and girls in an attempt to eliminate extramarital sexuality. The purpose is to construct a pure nation as mirrored by moral Chinese women. The superiority of the nation is maintained through valorizing Chinese moral values and heralding Chinese moral virtues as the most advanced weapon combating AIDS. The supremacy of the nation is confirmed and proved through virtuous women who are the bearers of the Chinese national essence. Women and youth are, hence, at the locus of education to safeguard their moral values and protect the purity of the nation, while men are exempt from the campaign's ideological controls.

Chapter 3 complicates the gender hierarchy implicit in practices of birth control and disease control by examining the state's restrictive attitude toward sex, manifested in the debate between vilifying and promoting condom use. The discussion of the condom debate helps situate the clients' attitude toward condom use in the restrictive political environment where illicit, extramarital, and premarital sex are targeted, regulated, and prohibited.

The next three chapters are purely ethnographic accounts, devoted to the clients' perceptions of condom use, hostesses' attitudes and strategies toward noncondom use, and clients' cultural views of sex and their control over sexual desires. Chapter 4 addresses the discrepancy between clients' knowledge about the efficacy of condoms and their refusal to use them. I argue that three major cultural and political factors are conducive to explaining the gap between understanding the protective value of condom use and the refusal to use condoms. These factors include (a) the rejection of state control over sexual expression, (b) the perception of contraceptive use as the woman's responsibility, and (c) the perception of nonuse of condoms as bravado (*meng*) as perceived by the client-peer group.

Chapter 5 addresses the hostesses' perceptions of condoms and how they respond to the clients' refusal to use condoms. Hostesses are forced to accept clients' refusal to use condoms to maintain their regular clients and glean financial benefits crucial to their livelihood. The costs of this refusal include repeated abortions and infections with STDs and high HIV risk. Hostesses exhibited little interest in learning about HIV because to them, surviving economically overrides health issues. To protect themselves against infection or pregnancy, hostesses availed themselves of presex antibiotic shots, emergency contraceptives, liquid condoms, external-use liquids, and napkins.

Chapter 6 employs open-ended interviews about male clients' sexual histories to discuss sexual matters and, in turn, the cultural context of HIV risk. I examine male clients' source and degree of sex education, their opinions about sex education, and their definitions of sex and control over sexual desires.

The conclusion makes intervention recommendations based on the theoretical and practical implications of my research on disease transmission and gender relations of power grounded in the everyday lives and worldviews of clients and hostesses. The implications are considered for the global transmission of HIV/AIDS and future development of HIV/AIDS research.

CHAPTER 1

GENDER AND PROPHYLACTIC
USE IN CHINESE HISTORY

A SCRUTINY OF THE IMPLICIT AND EXPLICIT cultural logic underlying gender dynamics and birth control is crucial to understanding the sexual inequality and condom use between male sex consumers and female sex workers. As stated in the introduction, the historical, social, and cultural regulations and obligations shape patterns of contraceptive use, negotiations of condom use in sexual encounters, and possibilities of sexual violence between men and women. What are these cultural rules and obligations concerning family planning and birth control, and how have they been formed throughout history? Who has been bearing the brunt of birth control? What kind of sexual parameters have the cultural rules set up for men and women, and how are they different from one another?

Answers to these questions will enhance our comprehension of the historically embedded cultural obstacles that obstruct women's demands for condom use. To achieve this goal, I will offer a historical account of how knowledge about prophylactics has been produced, constructed, and disseminated by the ancient imperial states, the Republican state, the Communist state and the post-Mao state, and how this knowledge has been learned and practiced by the Chinese people. I will demonstrate how family planning and priorities in, and prohibition of, prophylactics as part of the state's probirth, antigrowth, and birth control policy led to different constructions of the role of prophylactics and affected popular understandings of gender and the purpose of sex. This analysis also helps crystallize how sexual concepts and reproductive habits—often seen as inherently personal and therefore removed from the general stream of history—respond to broader social forces, such as state policy changes.

I will mainly draw on my research in historical archives; official and popular media, including popular and academic books, journals, magazines; and sex education CDs disseminated only to newly married couples; and my ethnographic interviews. I will unravel the historical continuity of women bearing the brunt of contraception throughout China's past and present. This account offers an excellent context for the readers to understand why many of the clients in my study, though in their forties and fifties, and married with children, have never or rarely used condoms throughout their adult lives.

ANCIENT AND LATE IMPERIAL CHINA

Confucian values, at least as they have been implemented in China, have always regarded females as subordinate instruments for the smooth functioning of a male-dominated family and state.[1] That this has been true throughout Chinese history is well illustrated by the following discussion of attitudes toward population control and sexuality in ancient China.

During this time, there were pronatal attitudes and concerns about the doctrine of filial piety and the fear of decreased population. It was believed that the central ideology of filial piety and the assurance of labor were crucial to long-term state security and a flourishing economy (Dikotter 1995). As Mencius said, "There are three things which are unfilial, but the most unfilial of these is to have no sons" (Z. Zhao 2006, 18). Being benevolent parents and filial children were heralded as the central ethics and Dao of the Heaven (*tiandao* means the way of nature that people have to observe as a model for social behavior;[2] G. Fan 2001), and large families were preferred to guarantee ancestor worshippers and old-age insurance (Himes 1963). Drowning and abandoning children was believed to be against the Dao of Heaven and disruptive of moral order. Between this entrenched ideology and the state's insecurity about the short life expectancy and high mortality of babies, a plethora of policies were implemented to encourage population increase, such as tax breaks, food aid, rewards of cloth and rice to large families, and punishment of those families that drowned babies (G. Fan 2001).

The pronatal attitudes and concerns during this period did not preclude a plethora of folk contraceptive methods that varied from forcing semen to flow backward, folk medicine, to infanticide. Although the Yuan dynasty prohibited prostitutes from abortion and the Ming and Qing dynasties applied heavy punishment to those who drowned babies, one of the means to control unwanted birth was still the drowning of babies (G. Fan 2001).

In general, sex in ancient China was distinguished by two separate purposes: sex for reproducing and sex for creating greater male potency[3] (Van Gulik 2003). The two fundamental principles of yin and yang in

Chinese philosophy are crucial to understand the meaning of sex. Yin conveys the meaning of being negative, dark, passive, cold, wet, and feminine, and yang carries the meaning of being positive, bright, active, dry, hot, and masculine. There is yin in yang, and there is yang in yin. Yin and yang interact with each other and influence people's health. Sex to enhance potency involved yang-strengthening and yin-replenishing. Yang-strengthening was accomplished through sexual arousal without ejaculation. The purpose was to cause a reverse flow of semen at the moment of climax rather than allowing the precious fluid to ejaculate (Van Gulik 2003).[4] It was believed that men could reach immortality through making the semen flow backward to be transformed from *jing* (seminal essence) to *qi* (vital energy) and then to *shen* (spirit) at the point when semen and yin coalesced in the brain. This process was completed through breathing techniques or pressing a particular acupoint. The importance of multiple sex partners was intended to enhance men's sexual prowess. Frequently men turned to concubines and prostitutes to replenish yang and to use the replenishment to impregnate their wives. Women's breast milk, saliva, and secretion were absorbed as the medicine for men to replenish their original Qi. Men were advised to absorb the maximum yin from the women by retaining their penis in the women's vagina as long as possible and appropriating the female red vital essence for themselves. Although a man must bring as much pleasure as he can to his female partner to absorb the maximum yin from her, his ultimate purpose is to attain his own immortality (Furth 1994; Van Gulik 2003). Therefore, I argue that a consistent theme was the appropriation by men of women's Qi for their personal immortality and for strengthening the state through procreating healthy children.[5] Interestingly, creating reverse flow of the semen was adopted by the Maoist and post-Maoist era as a contraceptive method (see F. Li 1958; Z. Zhang 2002).

Traditional Chinese medical texts did offer a myriad of contraceptives and abortifacients. For example, the rhythm method, that is, identification of the period of female fertility and avoidance of sexual activity during the time, was recorded during the Yuan dynasty. "Those who are afraid of too many births can calculate the dates . . . a woman can get pregnant during the first three to five days after the menstruation, with a son during the first day and with a daughter during the second day." The alleged safe period here was contrary to modern science, which could be used to explain the ineffective contraception during this time, and hence the unchecked population increase. Although the method was based on inaccurate science, it is possible that it expressed some concern for the health of women who have borne too many children. However, it might also represent the interests of some poor male peasants unable to support large families.

Because poor peasants desired to limit family size, the failures of contraception led to flourishing abortifacients during this time that were frequently harmful to women's health. Abortifacient prescriptions and contraceptive recipes were found in ancient medical texts such as *Qian Jin Fang* (*Thousands of Gold Prescriptions*) by Sun Simiao who died AD 695, *Yi Xin Fang* by Danbo Kanglai, *Furen Liangfang Daquan* (*Complete Collection of Valuable Prescriptions for Women*), and *Duanchan Fanglun* (*Contraceptive Prescriptions*; Himes 1964). It was recorded that an herb called *huirong* found on Fan Zhou Mountain could make a woman barren (G. Jiang 2003). Abortion formulas included ingredients such as barley leaven (*daqu*), liquor, silkworm eggs, thyme (*shexiang*), Chinese goldthread (*huanglian*), mercury, medicinal leeches (*shuizhi*), purple eggplant flower, rape seeds, cotton seeds, tadpoles, and so on (Himes 1963; G. Jiang 2003). Besides these recipes and formulas, it is reported that prostitutes applied disks of oiled paper to the cervix to prevent conception (Himes 1963). It was a piece of thin transparent paper, much like toilet paper, made of bamboo tissues, and was inserted into the vagina to prevent the penis from touching the uterus. Some inserted a piece of cotton, locks of hair, fiber, or silk into their vagina; others soaked the cotton in cooking oil, vinegar, or wine to increase their efficiency as contraceptives (G. Fan 2001). During the Maoist period, many of these abortifacients were revived and found to be seriously damaging to women's health.

REPUBLICAN ERA (1911–48)

The Republican government continued the pronatal policy. At the wake of the onslaught of Western colonial intrusion, the secure time of "culturalism" (Fitzgerald 1996) was superseded by the perception of the decline of China and the growth of Chinese nationalism. A new generation of intellectuals invoked the authority of medical science to replace Confucian philosophy to regulate sexuality and reproduction, as they were considered to be intimately linked to national strength and state power (Dikotter 1995). Government policy still insisted that the role of sex was to ensure procreation, rather than provide pleasure. Intellectuals argued that individuals had to stringently discipline and restrain their sexual pleasures because excessive intercourse retarded a man's sperm. Because of the belief that a woman carried all her previous sexual partners' semen in her blood, the importance of marrying a virgin was emphasized to preclude passing on the degenerate genes, and it was believed that excessive intercourse produced weak, malformed, or moronic offspring (Dikotter 1995).[6] On the basis of this consensus, intellectuals debated how sexuality and reproduction should be policed. The Guomindang government (Nationalist Party) adopted Sun Yatsen's pronatal ideas as the official ideology and favored an unconditional increase of population. Sun's three principles of the people were opposed to contraception and considered limitation on births a form of racial suicide. Birth control information was thus ignored by official publications and rarely discussed in government circles (Dikotter 1995).

Some intellectuals such as Pan Guangdan and Gao Xisheng proposed that birth control would lead to the decimation and extinction of the more gifted stocks in the population (Dikotter 1995). One can see in this the influence of the Western eugenics movement, which, along with many other Western ideas at this time, had a profound impact on China. Pan Guangdan expressed his concern about the lack of fertility among "the superior elements of the race" and suggested a birth release program for the higher strata of society so that the inferior elements would not swamp the professional classes (Dikotter 1995). Gao Xisheng, the author of a primer of birth control, also contended that 90 percent of the factory girls' progenies were mentally retarded (*dineng*), a dysgenic tendency on the rise because of the dissemination of contraceptive knowledge to the educated public (Dikotter 1995). Despite the absurdity of Pan's views, they were officially endorsed in the 1940s and resulted in the Ministry of Social Affairs organizing an unprecedented committee to study population policies in 1941. The committee recommended segregating those physically and mentally handicapped from the normal population and advocated a differential birthrate based on unequal birth endowments; that is, some individuals should have children while others should be sterilized to racially rejuvenate the country (Dikotter 1995).

Eugenics continues to influence contemporary Chinese's thinking about population. It is expressed primarily as a concern that women must conform to high traditional moral standards if they are to produce healthy and moral children, once again, making women responsible for the heavy burden of the nation's future.

The pronatal policy was modified by the concern of some intellectuals about eugenics. However, not all intellectuals accepted the pronatal policy. Influenced by Thomas Malthus's population principle that was introduced in China by a foreign missionary in 1880, proponents of reproductive control such as Wang Shiduo and Yan Fu proposed divergent views on limitation of births. Wang proposed drowning female children and physically abnormal or unattractive sons. He wanted to impose a tax on daughters to encourage infanticide. He also wanted to construct temples, nunneries, and halls of chastity to encourage celibacy. Anticipating the one-child policy, he even wanted to compel women who had already given birth to take abortifacient drugs. As shocking as infanticide is, it was a common practice in Europe and in Asia until the eighteenth century. Yan Fu advocated banning early marriage and a role for doctors in eliminating inferior babies. Proponents of reproductive control underscored the physical hardships, high mortality rate, and lack of hygienic care resulting from unrestricted procreation. A special issue of the women's journal argued that birth control would contribute to race regeneration (Dikotter 1995). While this period does show some uncharacteristic concern for women's health, ironically it solves the problem by advocating drowning and infanticide of female children.

Although abortion was not prohibited under Ming and Qing law, the early Republican Criminal Code (articles 332–38) proclaimed abortion illegal (Kotenev in Hershatter 1997). Similar to the situation of illegal infanticide during the pre-Republican era, criminal abortion during the Republican era was common in large cities, frequently leading to pelvic inflammation or death from sepsis. A plethora of drugs were sold as abortifacients; dangerous practices were also implemented, such as piercing the uterine wall with needles to induce abortion. Besides the previously mentioned abortifacients, abortion prescriptions from early Chinese medical texts were also used. Prostitutes during this era were given daily doses of alum or live tadpoles to eat to prevent pregnancy, induce abortions, or to cause infertility. It was believed that the "cold element" in tadpoles would counteract the "heat" of pregnancy.

During this era, the professional classes exclusively adopted contraceptive methods (Dikotter 1995). When it was discovered in a survey of thirty-four educated urban women conducted in 1930 that only three were informed about birth control, the birth control debate heated up but was always confined to the intellectual realm. In 1922, Hu Shi invited Margaret Sanger, an American crusader for birth control, to address the students at the Beijing National University. Students transcribed the notes of her talk and translated her book *Family Limitation*. Because of her influence, a twenty-six-page pamphlet was published in 1922, and a translation of her *What Every Girl Should Know* was published in 1925. In 1936, Sanger was invited again by Yang Chongrui, the president of the First National

Birth-Aid School, to deliver a speech in the auditorium of Xiehe Hospital in Beijing. Sanger's visit and the publication of her works invited a heated debate on birth control (Dikotter 1995). At the beginning of the 1930s, Yan Fuqing organized a birth control league in Shanghai; in 1936, Nie Kesheng established a eugenics society, mainly interested in contraception, in Hong Kong. During the mid-1930s, the popular press took up the issue and spread contraceptive knowledge in vernacular newspapers. Birth control pamphlets and handbooks flourished during the 1920s and enumerated all the methods available to the modern couple, even including directions about how to make condoms from animal guts.[7] A 1935 description of a condom (*guitou tao*) presented it as an instrument for enhancing male sexual pleasure, much like the penis ring (*yin tuozi*) of Ximen Qing in the novel *The Golden Lotus* (M. Yu 1935).

Although the Republican government policy was to prohibit birth control, as we have discussed in this section, because of the powerful Western influence of eugenics, American birth control crusader Margaret Sanger, and the dire prediction of Malthus, the policy was debated among a small circle of government officials and intellectuals. In spite of the policy and the heated debate, however, few ordinary Chinese were influenced either by the policy or by the debate. In most cases, they continued the ancient traditional practices of infanticide and abortifacients when they felt the need to control birth.

To clarify the meaning of this discussion for present-day China, it is important to understand that despite the presence of Margaret Sanger, the primary concern was not women's health, but the concern for overpopulation and the health of the patriarchal state.[8] Margaret Sanger was not concerned with overpopulation. She campaigned in China as she had in the United States for the rights of women. Her concern for women was the basis of her concern for birth control. In the United States, she witnessed poor immigrant women who carried the burden of ten to twelve children. For her, birth control became a key to the liberation of women, giving them more control over their lives. Unfortunately, in China when birth control was instituted during the Maoist era, this feminist concern was completely lost. In fact, as I will show, women were burdened by the one-child policy because of the way it was implemented.

MAOIST ERA (1949–78)

1949–52

The Communist regime advocated rapid population growth as a sign of prosperity and improvement in livelihood. In 1952, the newspaper *People's Daily* condemned birth control as killing the Chinese people without shedding blood. During this time, abortion and sterilization were strictly prohibited and controlled.

1953–56

The first national census in 1953 revealed a population of 602 million, much larger than the expected figure of 450 million (Liang and Lee 2006, 9). Alarmed by this large number, in 1953, Deng Xiaoping, then the secretary general of the Chinese Communist Party, wrote to Deng Yingchao,[9] emphasizing the importance of contraception. In 1954, Shao Lizi, one of the sixty-eight members of the Standing Committee, openly advocated birth control. At the first meeting of the People's Congress, he urged propagating knowledge about contraception and providing supplies of contraception.

At a ministry-level meeting in 1954, President Liu Shaoqi proclaimed the party's support for planned fertility, and in 1955, the party's Central Committee approved a regulation of birth report drafted by the Ministry of Health, which was known as the "Directive on the Problem of Population Control," pronouncing the first official population policy (Liang and Lee 2006, 10). In 1955, during the first National People's Congress, Premier Zhou Enlai also proposed birth control to protect women and children and to provide better for the younger generation (Tien 1973). This resulted in the first condom factory established in Guangzhou in 1955.

Also in 1955, the first article on contraception appeared in the official journal *Women of China* in response to an overwhelming number of letters requesting information about contraceptive methods (B. Dong 1955). The need for birth control is assessed through various testimonies from highly educated women who wrote to a journal to complain and were able to do so. This should be considered as a possible bias of the journal, not necessarily reflecting the general opinion. The editor Dong Bian defined contraception (*jieyu*) as the use of scientific methods to adjust the density of birth and contended that unscientific and unsanitary methods were dangerous to women, as many had already lost their lives from random abortifacients. She quoted some female cadre readers in their twenties who already had three to four children. They expressed their desperate need for birth control, as they were frustrated by their debilitating physical health and lack of energy and time for work. Thus, Dong invited Dr. Zhou E-fen to introduce scientific methods to direct the readers and stated that the purpose was to protect the health of women and children so that they could better serve the socialist construction, while caring for and educating the next generation. This purpose was proposed in tandem with Mao's leading principle, that is, to tap into the great resource of women in building socialism. Women were heralded as able to hold up half of the sky and thus were expected to devote themselves to social production. The editor, afraid of inviting potential criticisms, assured the readers that providing birth control information would not reduce the population, as only those who already had too many children would need birth control, and they could stop birth control if they wanted more children. After all, as the editor repeated, birth control was not an individual issue, but an issue concerned with national health and socialist construction, as it was indispensable that women were healthy builders of socialism. Although initially concern for women's health was discussed, in the long run it became clear that this concern was related to the larger issue of the health of the nation.

Dong also introduced the first two books on contraception and sex written by doctors, *Common Knowledge about Contraception* and *Sexual Knowledge*, published in 1955 and in 1956, respectively. The first book introduced the differences between birth control, abortion, and sterilization, and the second book discussed the biological and social aspects of sex, including contraception (Wang, Zhao, and Tan 1956; Zhao 1955).

Four themes stood out during this period: endorsing contraceptives, emphasizing sex for procreation, dispelling fears and misgivings about contraception, and coping with men's reluctance to use contraception. First, official sanction was given to the use of the rhythm method, condoms, diaphragms, cervical caps, vaginal tablets and suppositories, and jellies with or without diaphragm and cervical cap. However, condoms, cervical caps, and vaginal tablets were not only costly but also difficult to obtain. As recorded, one cervical cap cost three yuan (enough to buy fifteen *jin* of rice or four *jin* of pork at that time), and condoms could only be purchased through entrusting friends who traveled to Beijing or Shanghai at the cost of ten to twenty yuan for one trip (E. Zhou 1955). Given their high price and unavailability, some readers shared their methods of preserving condoms so well that one condom could be repeatedly used for a year without breaking.

Journal articles and publications discouraged sex during women's menstruation, the first three months of pregnancy, postpartum time, or after drinking because it would harm procreation (S. Wang 1956). Masturbation was heavily condemned as a crime and "a malicious habit" (*e xi*) to be overcome (G. Yu 1956). People with venereal diseases were prohibited from marrying according to the 1950 Marriage Law. Sex was restricted to and authorized only within marriage, and couples were advised to have intercourse no more than two times every week (F. Li 1958). It was emphasized repeatedly that human beings were "advanced animals" (*gaoji dongwu*) who had a natural desire to procreate. "It is one's heavenly duty (*tianzhi*) to be parents" (W. Wang 1957, 68). "The purpose [of sex] is not to satisfy instinctive desires but to pursue a beautiful life. If one does it only for sexual satisfaction, s/he is a low-level animal" (Wang, Zhao, and Tan 1956, 30).

Many journal articles and books attempted to dispel worries and doubts about contraception and sterilization (F. Li 1958, 48). One folk belief was that condoms debilitated health because they disrupted the coalescence and strengthening between yin and yang (Z. Gu 1956a). The folk understanding was that intercourse was to function as *yinyang xiangbu* (mutual replenishing of yin and yang); therefore, using condoms to disconnect yin from yang would dry the blood of men and women and make them frail, senile (*shouruo*), and neurasthenic (Z. Liang 1957). Some also feared that condoms would obstruct the smooth ejaculation of semen and harm men's health. Others believed that condoms would cause infection to the uterus. One journal article addressed each of these folk fears and claimed, "Today's doctors have not yet reached a consensus as to whether semen could be absorbed by the vagina membrane and whether the connection of the two could replenish each other" (Z. Gu 1956a, 26). Despite the author's ambivalence, it was clear that he had accepted the scientific view that rejected the yin-yang theory. The author then reassured the readers that condom use would not cause disease but excessive intercourse would. Thus, he warned against superfluous sex as too much ejaculation would affect the sperm count and lead to spiritual fatigue that could not be rehabilitated. Clearly, his adherence to science was subordinated to his adherence to the political dogma of restricting sex to procreation.

Journal articles also pointed out that pregnancies continued as a result of the lack of cooperation with the husband, as men felt condom use was too much trouble (*xian mafa, xian luosuo*; Z. Gu 1956a). For instance, in one letter, a female reader wrote that after giving birth to two children, she and her husband used condoms as a contraceptive. Both believed condom use harmed health; hence, they thought they had gotten thinner ever since using condoms. Her husband was averse (*dichu qingxu*) to condom use and complained about all the trouble. Afraid to strain their relationship, she submitted to his will and then became pregnant with the third child (Y. Wu 1956). Under such circumstances, some women wrote application letters for abortions and got approval from the working unit; others who were not approved because they had tried to induce too many abortions within one year resorted to other abortifacients such as deliberately jumping and playing basketball and squeezing the belly against heavy objects, which often led to massive hemorrhaging (Q. Ren 1997). Journal articles warned women not to depend on abortions for birth control. As of 1955, 70 percent of the women applying for abortions did not persist in contraceptive use (Z. Gu 1956b). This same theme reoccurred during the 1960s.

In 1956, the journal of *Women of China* published five herb prescriptions for contraception, collected from herb doctors and ancient Chinese medical texts (E. Zhou 1956). The article claimed that these prescriptions, tested and introduced by doctors, would not harm women's health, whereas others that had mercury, lead, and arsenic sulfide were poisonous. The wide publication of the herb prescriptions were in response to Shao Lizi's speech at the third session of the first National People's Congress when he applauded the Ministry of Health's attempt to collect contraceptive prescriptions from practitioners of traditional medicine. Shao specifically recommended one oral contraceptive formula of fresh tadpoles, washed clean in cold boiled water and swallowed whole three or four days after menstruation. If a woman swallowed fourteen live tadpoles on the first day and ten more on the following day, she would be barren for five years. She could repeat the formula afterward twice and be forever sterile. This formula originated with Yeh His-chun, lauded as an advanced herbalist by Shao. Tests of the formula under his direction started in March 1957 with the promise of abortions to the female volunteers, should pregnancies take place. Meanwhile tests on cats and white mice were carried out. Despite the heavier dosage than originally prescribed, 43 percent of the women became pregnant within four months, not mentioning the worries of their exposure to tapeworms and other parasites. In April 1958, tadpoles were officially declared to have no contraceptive value.

For the first time in Chinese history, traditional oral contraceptive recipes were extensively published in periodicals with nationwide circulation as well as provincial and municipal newspapers (Tien 1965). It was reported in 1955 that some five hundred thousand doctors practiced traditional medicine in China, whereas in 1958, only fifty thousand to seventy-five thousand doctors were trained in Western medicine (Tien 1965, 227). This revival of traditional medicine also included forcing semen to flow backward as a contraceptive method (F. Li 1958). Detailed descriptions taught men how to press a particular acupoint to make semen flow backward instead of discharging into the vagina. Both women and men were warned to be always vigilant for the moment to press. This method was categorized as withdrawal. Traditional medicine was promoted and gained renewed prominence in tandem with the official policy of cherishing China's medical heritage. Although tadpoles were officially disavowed as contraceptives in 1958, during the lapse of two years, there were reports from various parts of the country of deaths and disabilities from the ingestion of both specified and unnamed preparations. For instance, it was reported that some women took forty quinine tablets and became blind; some took safflower and went into a coma from hemorrhaging; some tore open the cervix with knitting needles, while others inserted corrosive herbs inside the vagina that caused serious infections (F. Li 1958). Other reports of prevalent side effects included stomach and abdominal pains, bodily aches, nausea, vomiting, diarrhea, and menstrual irregularities. A local clinic of traditional medicine in Henan province was accused of selling poisonous contraceptive powder made of calomel, leeches, and two other ingredients, leading to toxic symptoms among more than twenty cadres of a school, including swelling of the body, blistering of the tongue, blood in the uterus, and swelling of the vagina. Despite the reports, officials argued that oral recipes should not be rejected because incorrect or unsupervised use could have been responsible for the reported ill effects or pregnancies. After 1962, however, those who promoted preserving traditional contraceptives were publicly condemned (Tien 1965).

1957

Many hurdles obstructed birth control activities since its incipient stage. These obstacles included misinformation, doubts about contraceptives, men's reluctance to use contraceptives, and so on. Journal articles in 1957 continued to report women's frustrations and suffering from too many births. The All Women's Federation and Ministry of Health was inundated with letters seeking financial assistance and abortion information because of the birth of numerous children (J. Lei 1957). A female reader wrote to the journal *Women of China* discussing how she constantly invented excuses not to go home to avoid sex with her husband because she was afraid that pregnancy would disrupt her studies (D. Yu 1957). In the end, her husband was pushed further away from her. Other readers related how fear of pregnancy made them retreat from marriage (Xiongqiong Li 1957). From readers' reports, it seemed common that by age twenty-one, young married women were already mothering three or more children and had to leave school to take care of them (J. Lei 1957; Xiongqiong Li 1957; D. Yu 1957). Some decided to shoulder the double burden of study and childcare; others relied on their parents for childcare or applied for subsidies from the country (J. Lei 1957). In response to the overwhelming needs, another book on birth control was published in 1957. This book attended to both rural and urban women as it recommended that rural women use condoms, jellies, or salt mesh (*shiyan hu*—one spoon of salt plus ten spoons of water, plus a spoon of rice powder or starch; boil it into a paste, and rub the paste onto a piece of clean cloth, roll the cloth over, and insert it into the vagina).

After July, a political backlash brought serious repercussions to leading scholars who were birth control proponents. Ma Yin-chu was one of them. Ma, the president of Beijing University, joined the heated debate on the "population problem" and delivered a speech on the population question at the fourth session of the National People's Congress in 1957. His speech, titled "New Population Theory," emphasized that Malthus was correct in his view of the geometric increase of population, though he was wrong in maintaining that means of subsistence only increased in an arithmetic ratio. Ma saw population growth as the main obstacle to improved standards of living, industrialization, and elimination of unemployment and underemployment. Ma rejected the Marxist contention that population problems cannot exist in a socialist society and insisted that population be planned even in a socialist society. Orthodox Marxists severely criticized Ma as "adopting bourgeois Malthusian population theory" (Liang and Lee 2006, 10). His theory, according to his opponents, blamed the working-class victims rather than the capitalist system for creating their own difficulties (Liang 2006, 10). Criticisms of Ma were exacerbated by the Great Leap Forward and the Anti-Rightist political movement. Under Mao's slogan of cracking down on capitalist revisionists within the party, Ma was denounced as a political rightist and was removed from his post as president of Beijing University in 1960 (Tien 1973). During the Hundred Flowers campaign, a host of other proponents of birth control had also been accused of being anti-Community Party, antipeople, antisocialist, and antidemocratic dictatorship and of harboring political ambitions. Debates on the population problem were silenced not only by the Anti-Rightist movements but also by the economic disaster during 1959 through 1962.

1958–61

The birth control campaign was dropped during the Great Leap Forward and the institutionalization of communes. A vast population was regarded as an asset rather than a hindrance to economic development. "Hand theory" was popular at this time arguing that "one person has one mouth, but one person has two hands. Two hands can support five mouths." "People in the past worried about our overpopulation, but the idea has been overturned, the question is not so much over-population, as shortage of manpower."

Although birth control was resurrected in 1962, it was not entirely abandoned during this time. In the journal *Women of China*, female readers associated the use of contraceptives with the considerateness and thoughtfulness of their husbands (S. Chen 1958; Z. Liang 1957; J. Wen 1958; P. Yu 1957; S. Yu 1958). One reader praised her husband for being considerate (*titie*) because he consistently used a condom, and another reader insisted that her husband agree to use birth control after giving birth to their fourth child (Z. Liang 1957; S. Yu 1958). Other women were not as lucky. Letters related that some women were so strained with childcare that they spit out blood or quit study and stopped working. Because women were the ones burdened with heavy housework and childcare, men did not feel the need for birth control. Rather, they considered it troublesome and even obstructed women from using birth control (F. Li 1958). Hence, it became women's responsibility to persuade the men (S. Chen 1958). One article reported that, in a factory, women's committee members sent people to talk to the husbands who would not practice birth control. For instance, one woman had four abortions and was extremely weak, but her husband still refused birth control. After committee members "worked him" (*zuotade gongzuo*; talked to him to change his ideas) to take care of his wife's health, he finally realized that more abortions would induce danger to her health and therefore agreed to cooperate with her in birth control (J. Wen 1958).

1962–66

In 1962, the Central Committee and State Council issued "Directions on Advocating Family Planning" and thus resurrected birth control and emphasized leadership in family planning. The document listed family planning as an important government agenda and demanded dissemination of information, production of drugs, development of technology, and scientific research. Premier Zhou En-lai repeatedly espoused the significance of late marriage and contraception in developing the economy and enhancing population health and emphasized voluntary planned parenthood instead of coercion. In 1964, the State Council established a family planning committee with subdivisionary offices to take charge of the birth control business. Meanwhile a family planning professional unit was organized to coordinate scientific research. In 1964, the State Council added a budget item specifically devoted to family planning.

During this time, because the family planning campaign only centered on the cities, the countryside that composed 80 percent of the total population had not yet started family planning. Birth control information was only imparted to women who had more than three children and was unavailable to newly married couples and women with one to two children (B. Dong 1965). Condoms were not made available in the countryside, and people still carried doubts and fears about contraception and sterilization (Y. Zhang 1964). For instance, sterilization was compared to eunuchization and castration of a cock, and contraception was deemed inappropriate for newly married couples because it would deteriorate their relationships and affect their health (Anon. 1964; Z. Lei 1979; H. Xiao 1965). The result was the accelerating fertility rate from 6.18 percent to 6.26 percent between 1964 and 1966, far exceeding the rate of 6.1 percent during 1954 and 1958. Mao Zedong realized that birth control should be emphasized in the countryside. Because of mass fear of intra-uterine devices (IUDs) and other contraceptives (X. Yu 1965), in 1966, cadres, party secretary directors, party members, and All China's Women Federation members were the first ones to insert IUDs and accept tubal ligations and led others to do the same (B. Dong 1966a; Z. Jiang 1966; H. Liu 1966; L. Zhang 1966). In 1965, a Japanese condom production line was introduced in Tianjin to manufacture condoms (Anon. 2005b).

As mentioned previously, contraception information only targeted women, and the burden of family planning was fully laid upon women's shoulders (see also Andors 1983; Greenhalgh and Winkler 2005). It was emphasized that women should carry through family planning (B. Dong

1965, 1966b). This period continued the previous theme of men having no role in birth control. Female readers wrote about their experiences of many births because their husbands abjured condom use, complaining that it was too much trouble (*xian mafan*; Xu 1964). One woman related that after giving birth to two girls, she had to quit school and constantly felt exhausted (Z. Xi 1964). Her husband completely disregarded her complaints. After giving birth to the third child, she tried consulting her husband about adopting contraception, but he refused. She cried and commented that since men could not give birth, they could not experience the pains of giving birth and raising children, and therefore, they only thought about themselves. She tried her best to repress her anger with his insults and tried reasoning with him until he finally reluctantly conceded. It was not until after she became sick and he had to experience tiring childcare that he agreed to contraception. Another female writer resorted to a different approach (Geng 1964). She had three children and realized that men did not feel the same pressing (*poqie*) need for birth control. To change the situation, she took advantage of his rest time at home to talk to him, helping him to realize her painstaking efforts in childcare and to enlighten him because he "lacked real experiences (of the pain)." So she detailed how she was tied up with housework; how she suffered from a hemorrhage after giving birth that severely compromised her health; how she was exhausted from raising too many children; and how other families had problems due to too many children. The result was not complete success, as he still did not use precaution a couple of times and again impregnated her.

Thus female readers advised other "female comrades" not to be "soft-hearted" (*xinchang tairuan*) and submit to the needs of men (*qianjiu nanfang*). Rather, they should be "resolute" (*jianding*) because it was the female comrades' responsibility to shoulder the burden of pregnancy and childcare. Meanwhile, they insisted that male comrades who truly care for (*aihu*) their wives should be more considerate (*weiqizi shexiang*). Thus, the two indispensable elements persisting in birth control were (1) the wife should not submit to (*qianjiu*) the husband and (2) the husband should truly care for the wife, so he would not detest a slight effort (*buxian jushou zhilao*). The party hoped for success in birth control through determination and cooperation (Mo 1964).

1966–77

While the Chinese government endorsed birth control and planned population during the 1950s, the insufficiency of the public health system prevented the practice of birth control (Liang and Lee 2006, 11). Following a sharp population decline during the economic disaster of the Great Leap Forward from 1959 to 1961, 1962 through 1966 witnessed a precipitous population increase, with total fertility rates between 5 percent and 7 percent (Liang and Lee 2006, 11). The rapid population growth caused the government to proclaim a new directive advocating birth control and supervising the outcome of the birth control campaign (Liang and Lee 2006, 11). The government instituted medical research on contraception, trained health workers, and established family planning administrations on local levels (Liang and Lee 2006, 11).

Birth control activities were undermined and came to a halt during the Cultural Revolution of 1966. Population increased rapidly achieving its highest rate from 1967 through 1970, and the government did not resume the birth control campaign until 1970. This population policy was supported by Chairman Mao, who pronounced that "population must be controlled by all means" (Liang and Lee 2006, 14). His stand on birth control helped promote population policy among the public. During this time, because the policy of two children per couple every four to five years, unlike the one-child policy during the 1980s, was easy to follow, it did not encounter severe oppositions from rural areas (Liang and Lee 2006, 14).

In 1971, in response to the explosion in birth, Zhou En-lai convened a meeting to discuss family planning issues, and the result of the meeting, "A Report on How to Do Well in Family Planning," was forwarded to the State Council and was approved. The report established the fourth five-year plan, targeting the rate of population increase and proposed measures such as the dissemination of information about late marriage and family planning to every household. The report required improvements in the quality of abortions and sterilizations and called for strengthened research on contraception. It also demanded increased production and supply of a variety of contraceptive drugs and equipment for the countryside. Local governments, reproduction units, residential communities, birth control committees, public health personnel, and the network of rural health facilities were improved and more equitably spread throughout the countryside (T. 1977).

While during the 1960s, birth control centered on the cities, after the 1970s, emphasis shifted to the countryside (Andors 1983). A family planning team composed of "bare-footed doctors," female directors, and "sister-in-law team leaders" (*dasaozi duizhang*—an elderly and respectable woman) was formed. Communes trained some medical staff capable of doing simple contraceptive operations. The increased participation of women in the health care delivery system enhanced the level of health care available to women (Andors 1983). The country supplied all the equipment needed for abortions and IUDs. Cities organized medical teams to go to the countryside and propagate birth control knowledge. As a result of rural family planning, rural fertility rates dropped from 6.01 percent in 1971 to 2.97 percent in 1978. This result took several years of harsh policies to establish.

By the end of 1974, Mao Zedong reversed course and began emphasizing that "population must be controlled" and pushed further family planning in urban and rural areas. In that same year, fourteen different contraceptive tools, including birth control pills and condoms, were supplied free of charge to married couples by the hygiene department at the commune, street, county, and city levels. The statistics in 1971 showed 3.9 million abortions, 6.2 million IUD insertions, and 3 million sterilizations. It was reported that in the city of Tianjin, in 1978, 3,593 tubal ligations and only 175 vasectomies were performed. In the same city, the failure rate for a new kind of IUD, the stainless-steel double ring, was 13.2 per 100 women, of which 4.7 percent resulted from expulsion, 2.4 percent from pregnancy with IUD, and 6.1 percent from removal for complications (Ch'iu Lyle 1980). The prevalence of tubal ligations, a difficult and relatively dangerous operation compared with male vasectomy, indicates not only the male fear of diminished sexual capacity but also the pressure on women to assume responsibility for birth control.

In 1973, population quotas were formally included in the development plan for the national economy. At the end of that year, China proposed the policy of "late marriage, longer intervals between births, and fewer children." "Fewer children" was defined in terms of Zhou Enlai's model of "one is not too few, two are ideal, and three are too many." Two-child families were favored. In 1979, the number was dropped to one child per couple. It was the first time in Chinese history that family planning had been written into the constitution.

During the Maoist era, it was the socialist dogma that population control was necessary only in capitalist states rather than in socialist states. The early Mao's policy followed the socialist line that unlimited birth could be accommodated by an equitable distribution system. However, as the fertility rate in 1957 exceeded six children per family, it became apparent that, even in socialist China, the increase of population could not be accommodated. As the Maoist government attempted to address population control, it is noteworthy that even though the communist regime proclaimed the equality of women with the slogan that women hold up half the sky, the implementation of birth control policy fell mostly heavily on the Chinese women with frequently disastrous results. Despite the great control over the Chinese population exercised under Mao, it was not until Deng Xiaoping came to power that an effective birth control program was implemented.

POST-MAO ERA (1978–2006)

Although Mao died in September 1976, it was not until 1978 that Deng Xiaoping became the de facto leader of China. This led to a revolution as consequential as the Revolution of 1949, although with less violence. Deng Xiaoping was determined to be pragmatic rather than ideological and announced a new policy of market socialism. The new market economy and hence the consumer society influenced a change in the attitude toward sexuality. For some people, sex became an activity of pleasure, separated from love. The change in sexual behavior was not abrupt following the transition toward the market economy, and many couples, especially in rural areas, still tightly associated sexual activity with procreation (Liu et al. 1997); however, in urban areas, the change was marked. From this point on, discussion of population policy will occur in the context of Chinese commitment to the market economy. Following the rural decollectivization, agricultural production had increased rapidly. Family planning's objective was mainly to help modernize the economy and balance population and resources.

With the economic reform and open policy, the new Premier Zhao Ziyang projected a population of no more than 1.2 billion people by the end of the twentieth century (Liang and Lee 2006, 15). To accomplish this goal, in late 1979, the State Council proclaimed a one-child policy that restricted each family to only one child, except in the autonomous regions. Rural people resisted this compulsory policy because they depended on male labor in family farming. To soften the tension, in 1984, the government revised the policy drafted by the State Committee on Fertility Planning, which allowed families in rural areas to have a second child and even a third child under special circumstances (Liang and Lee 2006, 16).

In conjunction with the population policy, the revised 1980 Marriage Law raised the legal minimum age for marriage (to twenty-two for men and twenty for women) and made birth control a duty for both spouses. Couples were offered birth control information and sex education before marriage, and after marriage, one-child families received a one-child certificate that entitled both the child and parents to a host of benefits (Lang 1980). Noncompliers were fined or referred to the court if the fines were not paid. In the early 1990s, all social forces participated in imposing birth limits, which pressured cadres at all levels and encouraged massive misreporting, abuses, and coercions, such as locking people in offices until they complied, confiscating property for unpaid fines, and forced insertions of IUDs on widows (Greenhalgh and Winkler 2005; Liang and Xu 1997). Exacerbating this hard stance was the corruption of local cadres who misused birth revenues for personal consumption. From late 1982 to the late 1990s, the policy was "compulsory insertion of an IUD after the first child and compulsory sterilization of at least one member of a couple after the second" (Gao et al. 2002). Nie, in his work on abortion in China, contends that China's birth control program includes coercive abortion and that "a collectivist and statist morality is used to justify official perspectives on birth control, abortion, and fetal life" (M. Nie 2005, 64).[10]

As family planning only focused on provider-controlled methods of IUDs and sterilization, millions of women's health was harmed. The cheap stainless steel IUD caused anemia and led to uterine and vaginal bleeding, fever during menstruation, longer menstruation, and abdominal pain (Anon. 2006a). One-quarter to one-third of women experienced contraceptive failure and had to undergo abortions, which greatly affected their physical and mental health (Anon. 1991; S. Gu 1981). Results for noncompliers could be fatal. In a 1989to 1991 study, women carrying unauthorized pregnancies to term were four times more likely to die in childbirth than women with state-permitted pregnancies (Greenhalgh and Winkler 2005). The policy went so far as to designate the year in which a woman was allowed to have a child. In one case, a woman was forced to abort her child at the seventh month because her pregnancy took place in the wrong year. The operation left her crippled, and she almost died (Greenhalgh and Winkler 2005).

From 1998, the service providers were encouraged to attend to the clients' needs and offer more careful counseling on the risks and benefits of alternative methods of contraception. The 2002 law stipulated that contraceptives had to be "safe, effective and appropriate" but did not specify a particular method of birth control. The HIV/AIDS crisis added new duties to the population-and-birth system to educate the public and the name of condoms was changed from "contraception condoms" to "safe condoms" (Greenhalgh and Winkler 2005).

Despite the government's representation of the new program as an empowerment of women, in reality, the long-term emphasis on sterilization, IUD, and abortion had a profound effect on the women. In 1997, among the contraception methods used by 200 million couples, sterilization and IUDs composed 93 percent (80 percent in townships and 96 percent in the countryside), vasectomies made up only 9.2 percent, and condoms 4 percent (Cai 2000; Gao et al. 2002). Eighty percent of Chinese women have had at least one abortion, 9.9 percent have had more than three abortions, and some have had as many as twelve abortions (P. Wei 2005). In 1999, 45.55 percent of women employed IUDs and only 3.87 percent of women used condoms (Cai 2000). Indeed, popular media advised women to use IUDs after the child was forty-two days old and birth control pills after the child was one-year-old (Yao 2002).

My interviews with government officials at the municipal family planning office in Dalian also confirmed the emphasis on IUDs for contraception. In my interview with the local family planning department, the director said, "For married women, after having a child, we recommend IUD as the predominant form of contraception." Statistics from the State Family Planning Commission showed that among the married couples practicing family planning, only 4.2 percent resorted to condoms and 8.9 percent chose male sterilization. Therefore, women bore the brunt of contraceptive measures (C. Wen 2002).

Emphasis on female contraception bespeaks gender inequality. As mentioned, from the inauguration of birth control, women were heralded as the mainstay of the cause and assigned the heavy task of family planning (Liu, Zheng, and Wu 1979). Women models in family planning were set up to lead other women. It was women who had assumed the leadership and were also the major target of birth control (Kuang 1979). At the community level, birth control policy was enforced almost exclusively by women (Greenhalgh and Winkler 2005; Kuang 1979; Liu, Zheng, and Wu 1979). Family planning was understood to be a woman's issue, even though male vasectomy is a simple outpatient operation with no risk to the man, and tubal ligations for women is more serious and dangerous operation. Among 800 million operations from 1971 to 2001,

95 percent were performed on women (Greenhalgh and Winkler 2005). Such a "feminization of birth control surgery" was vividly explained by a male reader in the journal *Women in China*. As he wrote, his wife was a better candidate for the surgery because he was the household owner. If the operation did not go well, it would greatly impact the family. However, "women are after all women. If the operation harms their physical health to the extent that they cannot participate in production, it is OK because they can still cook and raise children at home" (C. He 1964, 30). Underlying his letter "lie[s] pervasive cultural attitudes affirming male superiority, male entitlement to sex, and male prerogatives in protecting the body from risk" (Greenhalgh and Winkler 2005).

Sigley (2001) also contends that even though family planning authorities have stressed that family planning is the responsibility of both husband and wife, the responsibility falls on the shoulders of women. A disproportionate amount of statistical revelation focuses on the reproductive practices of women when compared with men. As Handwerker (1995, 366) notes, invariably women are blamed for being both infertile and too fertile.

Condom use was represented as a contraceptive that reflected men's care and consideration for women because men actively (*zhudong*) took responsibility. Research in Greece has also revealed that women view men's responsibility in contraceptive practices as a signal of their love and caring (Paxson 2002, 320). Similarly, in China, nonusers were accused of placing their own pleasure ahead of the health of their wives. For instance, one male writer wrote that his wife had had an abortion because he abjured condom use (Anon. 2004b). After the abortion, his wife urged him to use the condom when he initiated sex, but, thinking that turning on the light was too much trouble and would destroy the moment, he pulled her back and sweetly coaxed her into submission. His nonuse led to another abortion, and this time she suffered much more than before. Because her health deteriorated after the two abortions, he advised her to have the IUD inserted, which led to fifty days of vaginal bleeding and waist pain. Every night the pain was so sharp that she tossed and turned and could not go to sleep. She started taking antibiotics and medication, but none stopped the bleeding. After a year, the IUD was expelled, so another was inserted. Insertions were repeated three times, and her suffering and bitterness were inexplicable. Only after these female contraceptive methods failed and her health was wrecked that he accepted condoms as a safer method (Anon. 2004b).

As illustrated in this male reader's letter, the husband's convenience, health, and well-being were more important than the wife's well-being. The effort to improve women's reproductive health and contraceptive choice during the early 2000s was stymied by husbands' unwillingness to use condoms, even when their wives' health was severely endangered and debilitated. Indeed, men's refusal to use condoms caused frequent unplanned pregnancies (Yao 2002; Zhang, Gao, and Tan 2002). Studies showed that abortions among Beijing women were attributed to their husbands' demand for sex and refusal to wear condoms (Greenhalgh and Winkler 2005). Men who used condoms either did so carelessly or used defective condoms. Women were advised to use postsexual contraception, such as taking emergency pills, inserting jelly, squatting or jumping after intercourse, or applying vinegar or soap water to a piece of cloth and inserting it into the vagina (Y. Yu 2003; Zhang and Lei 2001; Zhang 2002; Zhu and Li 1993). Besides these methods, an expensive "liquid condom" called *Kanglebao* for 38 yuan each was produced in 2002 and marketed as a new route to provide women with safety. It was marketed to help with women's reproductive health because unplanned pregnancies, STD, and AIDS transmission had become the greatest menace for women, and female contraceptives could not provide protection. Women's subjection to intercourse without condoms added to their chances contracting viruses. Although female condoms could free them from dependence on men and reduce the fear of pregnancy and STD infection, they were too complicated to use. Some women, especially the hostesses working in nightclubs and karaoke bars, welcomed the liquid condom because they believed that the ointment was convenient and effective in killing the sperm and STD and the AIDS virus and gave the women control over their health.[11]

Popular books sought to understand why men did not like condoms even although condoms could prevent diseases and were thin, light, transparent, and came in sexy colors and fragrances. To enhance excitement, condoms were also lined and creased to emulate the feeling of skin (M. Pan 2004). Chinese medicine was applied to some condoms in the belief that it increased pleasure and prolonged sex. Although popular books considered it the sexiest contraceptive method, men did not like it at all. Authors argued that boys might like condoms because they could prove their manhood and, at the same time, prevent pregnancy. But, as they became men, they would favor more direct intercourse, disregarding and ignoring women's demands. Condoms were thought to affect the quality of sex: "Condoms compromise sexual pleasure. Because they have a layer of latex membrane, they desensitized the male's penis considerably" (Yao 2002). Men reported that having sex with condoms was like "having a shower with a raincoat on" or "kissing with glasses on" (M. Pan 2004). Authors also pointed out that some men had difficulty maintaining their erection. Such episodes made them embarrassed and unhappy. As one author wrote, "men maintained that they did not feel like making love with the condoms on; condoms disrupted the normal sequence of sex life." When a man is sexually excited, tearing open a condom package, rolling the condom, and getting it in place makes it difficult to maintain an erection.

Of course, the popular media pointed out that it was natural for men to want progeny, and prophylactics prevented this. In China, women are required by the Marriage Law to submit to the sexual demands of their husbands because Chinese law does not recognize marital rape[12] (Cao 2004; Y. Gao 1998; Ji 2005). If a man refuses condoms, a woman cannot legally refuse to engage in sex, so she is forced to depend on the pill or IUDs that do not prevent viral infections (Anon. 2004b). In the survey, more than 80 percent of the women engaged in sex out of obligation to their husbands and only 40 percent of the men reported ever having talked to their wives about contraception (Anon. 2004a). There were even cases of wives killing their husbands because they had been continuously forced into intercourse (Y. Gao 1998). In a different case in Shanxi, a woman was tied up by her husband and her mother-in-law and beaten because she had refused to have sex with her husband (Y. Gao 1998). Clearly, Chinese women must contend with expectations that are not favorable to their health.

CONCLUSION

Prophylactics and other contraceptives in contemporary China reflect the historical continuity of gender hierarchy. Despite claims during the Maoist and post-Mao era about the empowerment of women, the reality is a continuing deep cultural bias against women that holds women responsible for birth control and the consequences of failed birth control. This history helps us understand the difficulty facing women in contemporary China who always bear responsibility for fertility and birth control in the service of the Chinese state and their male partners.

The historical continuity of sacrificing women's health for family planning still epitomizes the nature of family planning in the contemporary era. In the ancient era, female infanticide was commonly practiced during periods of stress such as famine. Even in the Republican era, a state policy of imposed female infanticide was proposed, and even today, as a result of the pressure of the one-child policy, female infanticide and abortion of female fetuses have become common again (Greenhalgh and Winkler 2005). During the Maoist era, despite the proclamation of gender equality, women were expected to be the leaders and the mainstay of population campaigns, and female contraceptive methods were emphasized frequently at the cost of women's health. Women were made responsible for implementing the population policy. For instance, traditional recipes that tremendously harmed women's health were propagated instead of male contraceptives such as condoms, and when it was deemed necessary to sterilize one spouse, almost always the more invasive tubal ligation was favored over less invasive male vasectomies. During the post-Mao era, for the first time, a coherent birth-limiting policy was propagated and effectively implemented, however, again, at a devastating cost to women. Female contraceptives such as IUDs and tubal ligations were favored over male condoms and vasectomies. The subordination of women was the constant theme. The post-Mao effective population policy was fulfilled only at the risk of women's health without significant sacrifice from men. Although at the beginning of 2000, we saw a shift of policy, purportedly favoring women's reproductive health and women's initiative in contraceptive choices, the entrenched cultural ideas of gender inequality and the consistently misogynist policies before 2000 continued to shape actual practices in China; that is, women continue to shoulder the responsibility of contraception. An example of this misogynist policy is the continuing existence of the law that allows a man to demand sex from his wife at any time and makes it illegal for a wife to refuse. One might argue that until a woman gains complete control over her body, she will not be able to share the reproductive responsibility with men and hence will not be an equal partner.

CHAPTER 2

NATIONALISM AND WOMEN IN THE DISCOURSE OF HIV/AIDS IN CHINA

INTRODUCTION

HIV/AIDS IS "AN EPIDEMIC OF SIGNIFICATION" (TRIECHLER 1987); yet, the discursive paradigm varies from society to society. In the West, including Australia, discourse on AIDS reflects a fear of homosexuality, prostitution, drugs, and deviance; xenophobia is also apparent (Lupton 1994). To contract AIDS is to be revealed as a member of a certain "high-risk group"—"a community of pariahs," who indulge in delinquent and deviant behaviors (Sontag 1989, 24–25).[1] Construction of AIDS as a disease of certain high-risk groups, rather than high-risk behaviors, has led to marginalization, essentialization, and stigmatization of the "deviant" groups as both the culprit and the object of blame (see also Porter 1997). In the early AIDS discourse in the United States, this construction generated a distinction between the disease's putative carriers and the "general population," identified as "white heterosexuals," to be protected from the contamination from the former groups (Sontag 1989, 27). Such a construction helps maintain the social hierarchy of the mainstream and the marginalized and leaves the risky behaviors of the general population unchallenged. It "secure[s] control by both governments and international agencies and 'encircle[s]' rather than empower[s] those who are its targets" (Hsu, Lin, and Wu 2004; Porter 1997, 230).

How is HIV/AIDS portrayed in the Chinese media? Who are the people on whom the blame falls? On which categories is the stigma inflicted? What are the cultural meanings in which HIV/AIDS is couched in the popular media, and for what purposes?

In this chapter, I will draw on official and popular Chinese media to examine the depiction of HIV/AIDS. In preparing for this research, I have scrutinized more than twenty popular journals and dozens of local and national newspapers over the past two decades, official journals such as the *Journal of Chinese Women* (*zhongguo funu*) from 1949 until 2006, and a myriad of Internet articles. The popular journals I have perused include *Family* (*jiating*), *Family Health* (*jiating baojian*), *Family Doctor* (*jiating yisheng*), *Bosom Friend* (*zhiyin*), *Metropolitan Women* (*dushi funu*), *Good Wife* (*hao zhufu*), *Marriage and Family* (*hunyin jiating*), *Girl's Friend* (*nu you*), and *Health and Life* (*jiankang yu shenghuo*). Among these journals, *Bosom Friend* and *Family* were rated as the most popular media in China by the 2004 and 2005 publication statistics (Anon. 2005a). In addition to written media, I also made field trips to universities to elicit survey answers and to watch campuswide HIV/AIDS education films with the students.

My findings on the popular and official media have demonstrated that the intersection of gender and nationalism is embedded in the discourse on HIV/AIDS in China.[2] In other words, discussing HIV/AIDS in China means discussing national superiority and women's responsibilities. Along with Emily Martin, I recognize that there are multiple discourses in a cultural field (1987). Jing Jun, for instance, has approached the construction of HIV/AIDS in the Chinese media from a different angle (J. Jing 2006). His article focuses on the incitement of fear and terror of AIDS patients in the media and the lack of social trust inherent in the processes of such representation (J. Jing 2006).[3] Jing contends that, while the Chinese government claims to have done a great deal to combat AIDS-related stigma, no officials have tried to examine and critique the news media's role in creating an environment of hostile public opinion to AIDS patients. As he points out, China encountered a multiplication of rumors of "AIDS criminals" in many cities by the end of 2005. "The progression of these rumors and associated panics indicates that China has a long way to go in combating social discrimination against AIDS patients" (J. Jing 2006, 168).

Since the emergence of AIDS in China, discourses have targeted foreigners and immoral Chinese women as vectors and transmitters of the disease. The boundary of the nation is secured through blaming foreign nations as the sources of diseases. The HIV/AIDS discourse allows the state to propagate a strict sexual morality in adult women and girls in an attempt to eliminate nonprocreational sexuality. The purpose is to administer a pure nation as mirrored by moral Chinese women, rather than controlling disease for public health. The superiority of the nation is maintained through valorizing Chinese moral values and heralding Chinese moral virtues as the most advanced weapon combating AIDS. The nation's supremacy is confirmed and proved through virtuous women who are bearers of the Chinese national essence. Hence, women and youth are at the locus of education to safeguard their moral values and protect the purity of the nation.

FEAR, TERROR, AND STIGMA IN THE MEDIA

Although the government designed a new five-year plan and established an AIDS prevention work committee in 2004 to combat AIDS (Jiansheng Liu 2004), the media were still infused with fear, terror, and stigma about HIV/AIDS.

Journal articles named AIDS as the "super cancer" and "century killer" that is "no less than nuclear terror" (C. Wang 2005; L. Zheng 2004). Hearsay spread that the virus could be transmitted through the air (Ziliang Liu 2004), and journal articles claimed that dental work, ear piercing, haircuts, teardrops, and contact with saliva could transmit AIDS (Bin 2003c; L. Xi 2005). It was reported that journalists and photographers refused to hold AIDS patients' hands or insisted on antibiotic injections before conducting interviews with AIDS patients. Police were also afraid to touch AIDS patients. There was no place to incarcerate AIDS patients. It was said that the government had to spend 70,000 yuan building a special jail for AIDS prisoners, which had to be sterilized each day (W. Wang 2005). The media posted pictures of police wearing insulated clothes, rubber gloves, and masks when watching, interrogating, or executing "AIDS criminals" (Du 2004).

As a result of the media's fear-mongering, ignorance about HIV prevailed. A public survey showed that 48.8 percent of the general population knew that casual contact would not transmit the HIV virus, 29.8 percent thought that toilets and swimming pools could cause infection, and only 33.3 percent knew that mosquitoes could not transmit the virus (F. Ling 2004).

Fear and ignorance fueled stigma against AIDS patients. Li Dun, professor of social policy at Qinghua University, dismissed those with AIDS infection as marginal people who did not make much contribution to the gross domestic product (Yang 2004a). He insisted that official statistics proved that for nineteen years, the number of infected and dead had been less than the total casualties from annual traffic accidents and less than the number of annual suicides. Therefore, he claimed, "In the issue of death, AIDS is not our priority." Heated debates in the Central Party Committee involved proposals from some leaders to quarantine AIDS patients and engage them in simple labor so that they could not transmit the disease to others (Jianqiang Liu 2005).

The media emphasized the moral implications of HIV/AIDS by focusing on "the greed, temptation, seduction, and desire that could be released from Pandora's box" (Bin 2003a). Those infected with AIDS were portrayed as morally delinquent, either pursuing sexual freedom or connected to crimes. The coined term "AIDS crime" flourished in the media when a proliferation of articles reported that some people infected with the AIDS virus used it as a threat to avoid arrest and incarceration. It was reported that infected individuals used contaminated blood as a weapon to commit random crimes such as stealing, raping, and blackmailing (Ai 2004; W. Wang 2005).

Presenting AIDS patients as "AIDS criminals," as Jing Jun contends, was based on "an extremely thin layer of evidence." It was "amplified, sensationalized and distorted to serve the purpose of criminalizing marginalized suffers of AIDS" (J. Jing 2006, 168). Jing argues that, despite the Chinese government's claim to have done a great deal to combat AIDS-related stigma, no officials have ever attempted to interfere with or critique the media's construction of hostility against AIDS patients.

Scholars have emphasized a considerable gap between official policies and actual practices (Micollier 2003). That is, although the Chinese government has promulgated laws against AIDS discrimination, neither authorities, health personnel, nor the general public accept the nondiscrimination law, and each national law prohibiting discrimination encounters a local regulation that contradicts it (Micollier 2003).

For instance, local regulations such as provincial laws prevent HIV-infected people from marrying, working, or swimming in public pools (Micollier 2003). It was argued that AIDS patients should not marry because they would bring disaster to society by transmitting the virus to hundreds of children. In one instance, a wedding banquet proposal by two AIDS patients was rejected by every hotel in a local area of Sichuan for fear of affecting business (Hu 2005).[4]

Because of the lack of confidentiality, doctors could potentially disclose their patients' HIV status to their employers. As Micollier notes, "Of the 10 principles on which there is consensus in China and which sum up the content of current medical ethics, not a single one concerns patient confidentiality or anonymity" (Micollier 2003, 37). Some hospitals did not allow patients to know the truth about their HIV status but found excuses to transfer the patients to another hospital (Luo 2005a). Other cases involved situations in which medical staff refused care for HIV-positive patients, fearing transmission of the virus (Micollier 2003). Cities such as Chengdu and Beijing have forced patients to take the HIV test before any surgery, and infected patients were refused treatment in the hospital (Du 2004; L. Xi 2005).

Local regulations also grant police the right to issue warnings to companies run by HIV-positive people and even seize their merchandise. Other stigmatizing instances involved reporters losing their jobs only because they reported a story about an HIV/AIDS patient (Tu 2001). A group of HIV-infected patients were evicted from a legally rented apartment, rejected by subsequent landlords, and forced to leave the district where they lived (Micollier 2003).

THE FOREIGNER AS TRANSMITTER

In China, the media portray AIDS as an imported product from the West. When the first AIDS case was reported in an American patient who died in Beijing in 1985, he was reviled as a foreigner who was alleged to have engaged in "abnormal" (homosexual) behaviors (Fang 1996). After he died, his room was disinfected for twenty-four hours, and his clothes, the nurses' clothes, and medical equipment were burned. He was hated and detested because he brought terror and fear to everyone in the hospital.

Thereafter, it was reported that foreign males had infected Chinese nationals in China and that Chinese immigrants were infected abroad (Fang 1996). Journal stories of the transmission of the HIV virus through tooth extraction and sewing machines in foreign countries emphasized that foreign countries were the sources of the virus. For instance, in 2006, a journal article reported that a returnee from abroad had been diagnosed as HIV-positive after arriving in Shenyang (Ge 2006). He claimed that he acquired the virus during a tooth extraction abroad. In 2001, a journal article recounted that another man had been infected while working in Thailand (Tu 2001). He said that a female colleague had cut her finger on the sewing machine. When he tried to help her, he accidentally cut his own finger on the same needle. His female colleague turned out to be an HIV virus carrier, and he was infected.

Since foreigners were identified as harboring "bad" and "abnormal" sexual values, AIDS was construed as mainly the problem of foreign countries and evidence of capitalist corruption and decadent lifestyles. Under the media influence, people commented that, since they could not go to the United States, they could never contract the virus, even if they tried to get it (Anon. 2006). Those infected are said to "lead a Western lifestyle" and exhibit weak will, poor self-discipline, poor morality, and a heavy reliance on drugs (M. Lin 2005). In discussions of sexual transmission, journal articles attribute HIV/AIDS to the bad and abnormal sexual practices of anal and oral sex (Bin 2003a; B. Wu 2004). When the sina.net Web site invited an HIV/AIDS expert to chat online, readers could not post their questions because words such as "oral," "anal" sex, or "intercourse" were forbidden. A computer window opened warning against using such words. Stifling communication makes it impossible to talk openly about AIDS issues (B. Wu 2004).

In 1987, the head of the Hygiene Ministry declared that AIDS could be controlled in China because the two means of transmission, gay relationships and promiscuity, were prohibited in China (Anon. 2006). After several years, the mayor of Kunming, Yunnan, reiterated the message in conferences that AIDS was the problem of foreigners. The official and popular discourse evinced a law on the mandatory testing of all foreign residents. Since 1992, national sentinel surveillance sites have been set up in many cities to screen the blood of foreigners, returnees from abroad, international marriage spouses, and other high-risk groups (J. Li 2000; Lin et al. 2000).[5] It was also considered indispensable to proceed with the most severe checks of foreigners at the airport. Surveillance stations were established at various airports to mandate AIDS virus tests for foreigners and returned workers from abroad (Anon. 2006; J. Dong 2005).

As Sontag observed, "It has been common to associate dreaded diseases with foreignness" (Sontag 1989, 54). China is no anomaly in this respect. In India, facing the onslaught of HIV infections at the end of the 1980s, the director general of the Indian government's Council for Medical Research declared, "This is a totally foreign disease, and the only way to stop its spread is to stop sexual contact between Indians and foreigners" (Sontag 1989, 80). Media in South Korea also blamed the AIDS menace on outsiders and the fear of foreign sexual contamination (Cheng 2005). Japan has construed AIDS as a foreign disease because it was introduced from the outside, primarily the United States, through infected blood products, and hence, they blamed foreigners, especially migrant sex workers, for sexual transmission of HIV (Buckley 1997).[6] Other countries such as Taiwan, Thailand, Nepal, Nepal, and the Philippines also echoed this pattern (Hsu, Lin, and Wu 2004; Pigg and Fraser 2001; Smyth 1998).

FOREIGN MEN AND CHINESE WOMEN

Foreign businessmen with decadent capitalist lifestyles and Chinese women with immoral values are deemed as the mediums of transmission (Chu and Shao 2005). The intersection between foreign men and Chinese women is a highly contentious issue in China. Brownell (2000, 227) has pointed out that, with the porous international boundaries as a result of the opening policy, "the gendered nature of international relations became increasingly obvious." The Chinese state has strived to police Chinese women's bodies to protect national integrity. For instance, the 1984 criminal law defined women who "seduce foreigners and have intercourse with foreigners" as "female hooligans."[7] In one popular movie, female hooligans who seduced and slept with foreign men were tracked down by government agents.

This first film on AIDS was titled *AIDS Patient* and was released in 1988. In this mystery, Tony, a foreign teacher at a university on the east coast, died of AIDS. The police organized an investigation unit to collect Tony's relics and track down the three Chinese females who had had sexual relationships with him. This investigating unit also included the female secretary from the foreign affairs department, Xiaoyu Wang. The unit screened the blood samples of all the teachers, staff, and students under the pretext of physical examinations. The result showed that two women were infected. The first infected female student was found immediately. The discovery made her desperate to commit suicide, but the police stopped her. She told the unit that Tony knew a woman at a restaurant, but it was impossible for the unit to locate this woman. The second woman infected had used a fake name, so the investigation was stranded. At this time, more relics of Tony's were shipped from abroad, including a picture and a shopping receipt from a Friendship store. From the picture, the unit located the woman who had married a rich Japanese businessman and had gone to Japan. The second piece, the shopping receipt, showed that Tony had bought a recording of *Madam Butterfly*. As it happened, Xiaoyu Wang, the female secretary in the foreign affairs department, also had the same record. Was it a coincidence? After a thorough investigation, the truth was revealed: Xiaoyu was the infected female they had been looking for. Xiaoyu came to a room at the seaside where she had had intercourse with Tony and committed suicide by setting herself afire.

The film targets women who might be attempted to have sexual relationships with foreigners and portrays how they are doomed because of their "immoral" behaviors. This cautionary tale is intended to warn women against intimacy with foreigners because foreigners will corrupt and infect them. While the mass media stresses that foreigners are the vectors of disease, it also shows the catastrophic consequences of women's immoral behavior.

The media not only warns women to stay away from the corruption of foreigners but also characterizes these women as instrumental and calculating. As the media show, it is these women's materialistic motivations and sexual drive that leads to their doom. Thus, it is women's responsibility to avoid being infected by foreigners. For instance, journal articles reported a twenty-two-year-old female sophomore student, Julia, who was infected by her foreign boyfriend (S. He 2005; J. Ma 2005). After Julia's foreign boyfriend left China to receive AIDS treatment, Julia was identified by his university foreign affairs department as his Chinese girlfriend. The party secretary called Julia for a talk. He asked her whether she had had sexual intercourse with her boyfriend, and she denied it. He told her that her boyfriend had AIDS, and he persuaded her to get tested at the Centers for Disease Control and Prevention (CDC). She tested positive.

Julia's online book titled *Diary of an AIDS Female University Student* (Ju 2006) received a variety of responses from readers. Some admired her courage, and others expressed their contempt. Comments varied from condemning her morality to calling for greater virtues among Chinese women.

Comments attacked her vanity and argued that she deserved the infection (*jiuyou ziqu*)—"she should drink up the bitter wine she herself brewed." Readers conjectured that she was a rural girl, and after entering the city university, she probably aspired to take advantage of any means to change her fate and get ahead of others. Her dream of using a foreign boyfriend to go abroad was "shrewd" and "stinky (*chou*)." Instead of meeting her goal, readers argued that she was "played (*wannong*) as a sexual object (*wanou*)." Readers wrote, "Don't put your shame here!!! You deserve it! You are fucking stupid [written in English]." Readers warned others that they were being used by the woman to make money so that she could cure her disease. Nonetheless, as commentators wrote, it was a good lesson for a vain and materialistic woman.

Readers also commented that lewd (*fangdang*) "whores" (*biaozi*) were the root of the HIV/AIDS problem in China. They believed that it served the women well to punish them, "The society would not be purified if such a woman were not punished. We hope those women who speak with the lower part of their body [vagina] can take it as a warning (*yinyiweijian*)."

Readers commented that foreign men in general are too lewd and Japanese men in particular are perverted (*biantai*). Therefore, "we should support home-produced (*guochan*) men, otherwise pure Chinese would become extinct. As long as she is Chinese, we should screw (*ding*) her before the foreigners do."

Readers also commented, "Let these horny and lewd (*fasao falang*) girls go to bed with foreign devils (*waiguoguizi*). Let's hope all of them get infected with HIV virus as soon as they go to bed with them. But never come to us Chinese men any more." Comments continued, "Chinese AIDS prevalence is caused by these women. After all, few foreign women came to China to cheat Chinese men. We hope Chinese women curb their behaviors, don't get too far . . . According to the Chinese traditional beautiful virtues, Chinese women should be reserved and conservative . . . They should respect themselves and love themselves."

Brownell (2000, 226–27) relates that during the time she was in Beijing as a foreign student she was aware of the intersection between gender and nationality in China. Chinese female students were warned not to visit foreign men more than once; yet, Chinese male students could visit foreign females many times. Outside of the university, Chinese women who visited foreign men in their hotel rooms were "frequently seized by security guards and detained for self-criticism" (Brownell 2000, 227). Brownell further points out that, Chinese male students were so angry with African students who had formed relationships with Chinese women that they staged an anti-African riot in Nanjing in late 1988.

While Chinese men demonstrate a paranoid possessiveness and protection of Chinese women, it is a patriotic duty for the Chinese man to have sexual relations with white foreign women (Barme 1995). In his article "To Screw Foreigners Is Patriotic: China's Avant-Garde Nationalist," Barme opens his article with an extremely popular scene in a wildly loved TV series *A Beijinger in New York* (*beijingren zai niuyue*). In this scene, Qiming Wang hired a white, blonde, and buxom New York prostitute. As he showered her with dollar bills, he demanded that she cry out over and over again, "I love you." The phrase "to screw foreigners is patriotic" originated from a Chinese immigrant to Australia, who believed that Chinese students and immigrants in Australia felt this way. If Chinese males feel patriotic about having sexual relations with foreign women, it would be considered traitorous for Chinese women to have sexual relations with foreign men (see also Brownell 2000, 229).

WOMEN AS TRANSMITTERS

Women with corrupted morals were depicted as the major source of HIV/AIDS transmission. Categorized as "female sex criminals" (*xing-zuicuo funu*) or "clandestine prostitutes," they were reported as "the most dangerous group" that spread the disease to the general population. A proliferation of journal articles illustrated how the infected prostitutes endangered society by deliberately transmitting their disease to the general public, although the results of the national sentinel surveillance surveys continued to show that HIV infection prevalence was low in these selected populations (Qu 1997).

The general public as well as some professional health workers, doctors, and intellectuals held this belief. When professor of anthropology Jing Jun at Qinghua University entered the Beijing testing and consultation center, he expressed concern that he might be infected. The staff member immediately responded, "You don't look like a drug addict. Stay away from the hostesses in the future" (Luo 2005b). Jing Jun surmised that apparently, the health worker believed that unless the person was a drug addict, hostesses were the agents of virus transmission.

Some doctors shared the same belief. It was reported that, in an AIDS media training class run by Qinghua University and Sino-Britain STI and AIDS prevention project, trainees were asked, if they had the magic medicine for AIDS, who should they give it to—a long-distance driver infected by prostitutes, the driver's housewife, a prostitute, a patient infected by blood transfusion, or a homosexual male? Over half of the trainees chose the housewife, apparently because they were concerned that she was an innocent victim; yet, Dr. Chen Liang from the Sino-Britain STI and AIDS prevention project thought strategically about who would be more likely to spread the disease and chose the prostitute.

In this training class, both the doctor and trainees perceived the prostitute to be the agent rather than the victim. What differentiates the doctor from the trainees was the inability of the trainees to think strategically about what agents would be more likely to spread the disease. This rhetoric is also embedded in the state prevention policies, as the state designates prostitutes as the high-risk group and target of surveillance and investigation (Bin 2003a).[8] Women do not commonly transmit AIDS through sexual contact with other women. As more infection is identified among underground prostitutes (Q. Zhang 2004), a narrative that fails to hold men at least partially accountable for sex-induced increases in female

AIDS is dishonest. This rhetoric portrays men as victims and women as perpetrators.

Journal articles and films portray how the lives of vibrant and talented young men are ruined and atrophied by hostesses, known as clandestine prostitutes. One journal article reported on a young man in his early twenties (Bin 2003a). After he graduated from college, he went out drinking and visited a hair salon, where he had sex for the first time. This encounter infected him with the AIDS virus. According to the article, "He is still so young that he can run across the street congested with rumbling cars. He loves to laugh—the kind of very affectionate and intimate laugh." By emphasizing the young man's energy and promise, the article intends to trigger the audience's sympathy toward him and resentment and hatred toward the perpetrator—the hostess. Indeed, mass media are replete with stories of how hostesses seduce and infect male college students and successful men with sexually transmitted diseases (STDs) and HIV/AIDS (see Guan 2001; Yaling Li 1999; Jianlin Liu 1999; W. Shen 2001). Newspaper articles describe hostesses as "shrouded by death" because they are the vectors of STDs and HIV/AIDS viruses (L. Xiao 2001). Deeming sex workers as perpetrators of sexual transmission of HIV resonates with the media in other countries such as Japan, Thailand, and the Philippines, where sex workers are required to undergo mandatory blood screening for HIV (Law 2000).[9]

Hostesses as well as women with corrupt morality are portrayed as agents of transmission. The first youth AIDS education film titled *Youthful Repentance* (*Qingchun de Chanhui*) was shown at Beijing University in 2005. The film narrates the story of a handsome and vibrant male college student Shan Huang who is infected by Li Zhang, a female manager of a hotel where Huang interns. When Huang began his internship in Zhang's hotel, Zhang became sexually interested in him. She noticed that Huang wished to retain the job after graduation, so she used her power to promise him the security of the job. As a return for her favor, Huang followed her to the seaside and had sex with her in her car. Later, when Zhang was diagnosed HIV-positive, she informed Huang, who in turn tested seropositive. This dialogue basically reverses the customary male-female role, turning the women into the aggressor and male into the seduced sexual object.

During the film, we are never told how Zhang is infected, and no one realizes that she is a victim because she has been infected by a man. Instead, we learn that she is an evil perpetrator who uses her power and wealth to corrupt a youthful, bright, and energetic college man. We see the young man riding a bicycle by the seaside at sunset; we see him embracing many colorful wishes for the future after graduation; we see him enjoying his romantic relationship with a pure and loyal college girlfriend. While the young man is in his early twenties, Zhang is in her late thirties, impure, wealthy, and seductive. She lures Huang with power and wealth into having sex with her. She is the corruptor, transmitter of diseases, but also "the third party"—a term used to refer to women who separate happy couples and destabilize families.

The film affirms that women like Zhang should be abandoned and punished by the society. In the film, Zhang's attempts to redeem her conscience are ruthlessly rejected by everyone she has wronged. When she told Huang that she loved him and had not meant to hurt him, Huang yelled, "I was sentenced to death by you!" When she offered Huang's parents 600,000 yuan and apologized to them, she was driven out with a slammed door behind her and a curse, "you shameless woman" (*choubuy-aolian*). When she asked Huang's girlfriend's forgiveness, her roommates called her "a bully." "Why don't you find a man yourself? Stop seducing other women's boyfriends!"

Although Zhang is a morally corrupt woman, a perpetrator, and a third party, she is induced by her HIV/AIDS status to do good deeds and seek forgiveness from her "victims." She is deservingly discarded by everyone and society. We as an audience are led to share the victims' distress over Huang's tragedy and resentment toward the female perpetrator, Zhang.

There is no interest in the media to discuss whether immoral women like Zhang or, for that matter, any women could be potentially put at risk of infection in their private lives. Rather, they are part of the discourse of blame and viewed as responsible for transmitting the virus to men.

WHY WOMEN?

Historically in China, women's bodies have been understood as dangerous and dirty. For instance, during the Republican era (1911–48), a woman was believed to carry all of her previous sexual partners' semen in her blood, hence marrying a virgin was emphasized to preclude passing on the degenerate genes of former partners (Dikotter 1995).[10] Marrying nonvirgins ran the risk of latent sperm from previous lovers corrupting the blood line. This belief led to policing women's sexuality. Furthermore, studies have also shown that traditional Chinese medicine and popular Buddhist teachings identified women's menstrual bodies as polluting and dangerous (Furth 1992). Chinese traditional medicine identifies the menstrual period as a time of dangerous vulnerability and prescribes herbal tonics for menstrual health. Buddhist teachings focus on the polluting menstrual blood. Women were believed to have the capacity to use their menstrual "dirt" or "poison" to harm men and the agnatic descent group by causing illness or ill luck (Wolf 1972; Furth 1992). The purpose of Chinese medicine is to regulate women's menstrual periods to produce a normal and healthy body. While the beliefs that women's bodies are dangerous and dirty are not unique to China, they are still historically important in helping us understand the contemporary emphasis on controlling women and their sexuality.

In his analysis of the history of sexually transmitted diseases in China, Frank Dikotter observes that, "the regulation of sexuality, rather than the control of disease, has been the main objective of medical circles from the late imperial period to the People's Republic" (Dikotter 2004).[11] He states that economic and social changes along the coastal regions during imperial China generated heightened concerns over regulation of sexuality and the emphasis on the nuclear family among gentry scholars (Dikotter 2004). This theme recurs in contemporary China. Traumatic social and cultural changes frighten people, who respond to their fear by attempting to control the nation's morality. This response is represented by a renewed concern for control over women and their sexuality.

Chinese intellectuals have argued that China is experiencing a separation of sex, love, reproduction, and marriage (A. Wang 1995). The first, they claim, is the separation of reproduction and marriage. Marriage was transformed from reproduction of descendants to conjugal relationships that emphasize love rather than lineage. The second separation, as they argue, is between love and marriage. Extramarital affairs have become prevalent, which has destabilized marriage and broken the association of love, sex, and marriage. The third, they observe, is the separation of sex and reproduction. As a result of the technology of contraception and abortion, people associate sex with pleasure and favor nonprocreative sex. The fourth separation, as they contend, is between sex and love. Sex can be conducted and satisfied outside of marriage and love. The commodification of sex alienated sex from love and led to casual sex devoid of love.

These four layers of separation are also described as "four sexual revolutions" that have taken place in China (Z. Li 2005). As sexologists in China argue, the first sexual revolution took place in the 1980s, when sex, because of the one-child policy, was no longer for procreation. The second sexual revolution peaked at the beginning of the 1990s, when people started considering sex as a right. The third sexual revolution occurred in the mid-1990s, when sexual experiences were no longer limited to married couples. The fourth sexual revolution rejects monogamy of both marriage and cohabitation and includes multiple liaisons and even "wife swapping" (J. Nie 2004).

This four-layered separation, or four sexual revolutions, disrupted social morality and challenged marriage. To save marriage and family, the media have laid the responsibility on women and pressured women to defend social ethics. The *Journal of Women of China* published "Marriage: Women's Rights and Responsibility" and four installments titled "Women Are the Last Defenders of Ethics" (Liu and Xu 2005). In the first article, two female writers discussed the responsibility of women in a marriage, that is, the responsibility to give up some part of their selves and make certain concessions to their husbands and their responsibility to take care of their husbands. They argue that failure to do so will lead to marriage dissolution.

The prelude of the four installments of women as the "last defenders of ethics" states that family is the most basic cell of society (L. Wen 2005c). In the midst of moral decay and a crisis in trust, the editor contends that women as wives and mothers should shoulder the responsibility of defending ethics in the family. The first installment discusses how women should be the first engineer of their children's souls, to educate their children to become kind, helpful, and sympathetic persons in society (L. Wen 2005c). The second installment advocates that women should "forgo individual but selfish rights for the sake of society" (L. Wen 2005d). These desires, as the article contends, can either betray one's basic conscience or betray public morality. A female reader wrote that she used to pursue material gains and luxurious lifestyles but found herself feeling more depressed and empty. In the end, she decided to retreat to the countryside and teach rural students. Previously, she had indulged in a sea of individual desires and had been tempted and dictated by fashion and modes. Now she was satisfied with the peaceful campus and the cramped dorm in which she lived. Another female reader wrote that women should explore and develop (*fayang*) their innate (*tiansheng*) empathy and kindness in society and not fall into the abyss of immorality. A third female reader wrote how she clings to her conscience (*shouzhu benfen*) and does not resort to bribery when facing competition from her female classmate for the opportunity to attend a conference in Hong Kong with her advisor. The third installment points out that "a husband would not be corrupted if a family had a reasonable wife." It emphasizes the key role of women in the "basic tune of the family" (*nuren goucheng jiatingde jidiao*) and claims that wives should take responsibility for their husbands' corruption because wives pressure husbands to strive for wealth and material goods (L. Wen 2005a). In fact, these articles demand that wives of court cadres should supervise the court cadres' "life circle, social circle and entertainment circle." In one instance, wives were made to sign a contract with the court to assume the responsibility of supervising the court cadres after work (Dong and Wang 2003). "How can anti-corruption have nothing to do with women? In reality, don't we see the phenomenon of 'wife is blowing the wind beside the pillow, and husband is corrupt and jailed?' Don't we see the phenomenon of the conniving wife indifferent to her husband's corrupt activities?" Since the wife started supervising the husband's activities, it was reported that the husband reduced expenditures on entertainment (*chihewanle*; including sexual encounters). There is no evidence that male behaviors have changed, only the claim made by the article.

The fourth installment similarly argues that "if every family is a clean land, then our society will have a clear sky" (L. Wen 2005b). It emphasizes that family is the most fundamental cell of a society and will transmit its morality to the society and, at the same time, reflect the morality of a society. Hence, family and society belong to a circular system that shares the same morality. The articles listed examples of wives engaging in illegal transactions and accepting briberies without their husbands' knowledge. Hence, the articles advocate that women should act as "virtuous wives and wise mothers" (*xianqiliangmu*), manage clean accounts in the family, and not engage in bribery or embezzlement. Instead, they should shoulder the burdens in the family such as housework and cooking.

In tandem with the proclaimed responsibility of women to keep family and society clean, articles also point out that women are responsible when husbands have extramarital affairs. First, these articles claim that husbands have extramarital affairs because their wives lack sexual interest (M. He 2003). The husbands are not satisfied sexually because women are submerged in trivial affairs, which makes the husbands look for spiritual comfort outside. The articles stated that extramarital lovers tend to be sexier, while offering them more comfort and satisfaction than their wives. Therefore, if wives can provide physical and psychological satisfaction, men would not betray them. Both Gary Sigley and Harriet Evans have argued that in China it is women's responsibility to manage family affairs, and sex is one of them. The burden of couple's harmonious sex is placed on women, who are instructed to bring sexual pleasure to their husbands (Evans 2002; Sigley 2001). Sigley contends that in China women must control their own physiological processes, be aware of the desires of their husbands, and satisfy and temper them. Hence, as she has observed, "much of the call for the emancipation of women in China has meant not casting off the shackles of marriage and the family but, rather, reconfiguring the lines of power within the family so that women become the primary conduits for government's programming. In this case the state-family relationship is restabilized with the mother as the main relay for adjusting certain reproductive and sexual practices" (Sigley 2001, 134).

Other journal articles narrated stories of wives successfully correcting their own mistakes and retrieving their husbands after they had gone astray. A female writer recounted that she forgave her husband's extra-marital affairs and admitted that she had ignored her husband after she had given birth (T. Zhao 2002). She realized that "he needs to be taken care of," so she adjusted her own lifestyle and squeezed out a half-hour to accompany him every day, no matter how busy she was. She not only decorated and cleaned the house but also dressed up beautifully. She bought sex science books to read with him so that he could enjoy sexual happiness and satisfaction. She lamented that if she had behaved this way before, his extramarital affairs would never have happened.

In another article, a female reader wrote that her husband had extra-marital affairs and proposed divorce (Yali Li 2002). Although her husband lived with the third party, she continued waiting for her husband to come back. She reflected on her own mistakes in dealing with family problems and actively offered financial help to him when he was in debt. Three years later, he returned home. Similarly, other wives enticed their estranged husbands to return home by bailing them out of trouble (Guo 2002). Other stories also pointed out that when the wife is not gentle and womanly enough, men may turn to extramarital affairs (Y. Ling 2003).

The journal points out that many wives worry about marital changes, so they start a long-term marriage protection war to keep their husbands and to secure the family (Y. Ling 2003). The journal publishes these stories of how wives secure their marriages as lessons to their female readers. The journal indicates that wives should forgive their husbands' extramarital affairs, even when husbands are caught by police sleeping with prostitutes (G. Jin 2003). According to the journal, these stories teach that, a wife, facing her husband's change of heart, should reflect on her own shortcomings and forgive her husband.

Media portrayals exacerbate the pressure on women to take responsibility for keeping the family intact and for securing social morality. Pressure also manifests in the media's fixation on teenage women and their strong criticisms of female college students.

Mass media criticizes female college students for their loose attitude toward sex, their acceptance of cohabitation, and their high abortion rate (Chu and Shao 2005; Zengqing Li 2003; H. Zhou 2005). Newspaper articles focus on the aberrant behaviors of female college students rather than male college students. Stories about female college students caught having sex on campus at night, who cohabit with boyfriends, or who act as bar hostesses in nightclubs are common (Zhen Liu 2002; You 2000). Articles point out that these women are brought down by the new capitalist economy and their pursuit of sensual pleasure. Female college students are portrayed as slaves of sexual desire, pursuing material enjoyment and conspicuous consumption of fancy clothes and fashionable accoutrements. Articles argue that some of these women fall into a "criminal abyss" because they treat sex as no different from satisfying hunger. Women are educated to remain chaste despite men's sweet talk and boyfriends' relentless sexual requests and advances (Yuan and Yun 1990).

Female college students responded to the editorial board, criticizing articles that blamed females for their sexual behaviors and ignored the responsibilities of men. Their letter insisted that the pressure females faced was greater than the pressures faced by men and that the psychological burden brought by public opinion was even heavier. They claimed that the articles exhibited a resurrected ancient demand on women.

Despite female students' efforts to strike back, social pressure on women increased. An example of this pressure was the so-called New-Chastity Campaign in universities. This chastity campaign was more an influence from American Christian group's abstinence campaigns than the resurrection of Chinese tradition (Zhen and Longyin 2003). It was reported in Sichuan University that eight female college students had signed a contract refusing sex during their four years in college and had agreed to supervise each other. This "virginal alliance," they contended, was meant to protect themselves from STDs and pregnancy because they believed that STDs were inflicted on the promiscuous. So the chastity campaign would prevent them from contracting AIDS. This premarital asceticism, according to the article, expressed young people's yearning for a traditional family and marriage.

Students in other universities such as Xiehe Medical University, Politics and Law University, Capital Normal, and Beijing Normal University also promised to remain chaste. Students in these universities have been part of an international education fund financed by American Christian churches that since 1994 had trained them to promise "youth abstinence" (*qingchun chunjie*; G. Gao 2000). At the end of the training, all students promised their devotion to youth abstinence and marriage harmony.

College students wrote articles that exhibited a nationalist pride. They deplored the Western crisis of immorality that generated promiscuity, pregnancy, STDs and AIDS, drugs, violence, and crime and condemned the "flies from outside that destroyed our environment, corrupted our thoughts, and disrupted our traditional morality" (G. Gao 2000, 42). Articles criticized American sexual liberation during the 1960s and 1970s and condemned the severe consequence such as rape, high divorce rates, and child abuse. Compared with America and other Western countries, articles heralded China as having much lower divorce rates, fewer teenager pregnancies, fewer single-mother families, and fewer AIDS patients (G. Gao 2000, 159).

Therefore, articles proudly argued that China should absorb and develop the essence of Confucian thought, rejecting sex education in favor of moral education (G. Gao 2000, 40). China's superior national characteristics, they argued, would allow them to do better than Western countries (G. Gao 2000, 58). After all, "we grow up in a country with a 2,500-year history of Confucian culture, a culture respected by the whole world" (G. Gao 2000, 40). "Given our better thought and better moral foundation, we should take advantage of our national essence and continue moral emphasis on family, school and society. Furthermore, we should abandon the foreign waste (zaopo) and maintain our own characteristics and our excellent tradition" (G. Gao 2000, 23). The articles insisted that China would flourish because of its "spiritual civilization" and its "excellent cultural heritage as a world-renowned civilization" (G. Gao 2000, 160).

Despite the relative lack of understanding of the nuances of Confucians among Westerners, these articles claimed that China's traditional moral values were now leading the world (G. Gao 2000, 94). Articles called on the Chinese people to defend themselves when facing the onslaught and infiltration of the Western "uncivilized and unhealthy lifestyle" and ensure that they remained clean and spotless. Only in this way, as the articles argued, could we create a "healthy and brilliant spiritual civilization" and use our Chinese traditional morality to influence the world. "Soon in the future, our great country must be the first country to drive away AIDS" (G. Gao 2000, 94).

Sontag observes, "AIDS is a favorite concern of those who translate their political agenda into questions of group psychology: of national self-esteem and self-confidence" (Sontag 1989, 63). This national self-esteem and self-confidence is embodied in abstinence and self-love (*jieshen ziai*) as the most effective weapon to safeguard against HIV (Bin 2003a; Y. Chen 2005; Xie 1997).[12] Articles advocate sex education emphasizing the danger of sexual freedom and the importance of abstinence and insist that women should be warned of the danger of frivolous sexual conduct and educated to treasure self-respect and self-esteem (Bin 2003b).[13]

Respondents specifically emphasized that the pressure to maintain such moral discipline should remain with women more so than men and that girls should reserve their chastity and not "degrade their humanity" (G. Gao 2000, 29). Articles criticizing women who had lost their self-esteem by looking for millionaire boyfriends stressed the importance of self-love (G. Gao 2000, 74).

Journal articles related some training lessons for female teenagers. For instance, in Chongqing of Sichuan, doctors in City Family Planning Hospital brought in forty-seven female students from nine to seventeen years old, accompanied by their parents, to visit the abortion operation rooms (Qu 2005). They intended to educate the young girls by exposing them to the harsh reality of abortion. They told the female students that, "if you engage in premature intimate contact with men, one day you may have to suffer the damage of these operation." Doctors explained in detail how sharp operation instruments are inserted into the uterus and then sucked out the fetus. After the visit, one girl claimed, "I do not want ever to enter this room again in my life. After school starts, I will tell my classmates about it so that the girls will know how to love and respect themselves [*zizunziai*]." Some parents considered it a violent education that extinguished the girls' rosy hope for the future. According to them, these terrible scenes were cruel for adults, let alone for girls. They argued that teenage girls should not have reproductive education as their psychological state of mind is not mature, and it potentially may generate negative feelings toward future sex life, making them not want to get married at all.

That education and training centers on young female students and not male students shows that women are used as the barricade against HIV transmission. Women are made to shoulder the responsibility of keeping the society clean, embodying and demonstrating the superiority of the nation. They are the weapons the state uses to fight HIV/AIDS.

Professor of sociology Pan Suiming at Renmin University echoes the claim that women are the barricade against the spread of HIV/AIDS (Yang 2004b). According to him, Chinese women are the reason that the projected 10 million cases of AIDS have not happened. Because most Chinese women have been faithful, AIDS has not been transmitted on a large scale. As he indicates, if a man goes to a hostess and gets infected, he gives it to his wife and the transmission chain ends there because the wife is faithful.

CONCLUSION

I have argued that nationalism and male dominance underlie the discourse of HIV/AIDS. National superiority is confirmed in condemning the infiltration of foreign corruptions and identifying foreigners as the sources and transmitters of the HIV virus. National pride is heralded in the Confucian morality and chastity deemed as the most effective and superior Chinese weapon to fight HIV/AIDS. National purity is presented as embodied by virtuous Chinese women.

Blaming foreigners and immoral Chinese women as vectors and transmitters of HIV/AIDS confirms Paul Farmer's theory of "geography of blame" and Carol Vance and Leclerc-Madlala's argument about the danger of women's sexuality and women as the agents of pollution (Farmer et al. 1993; Leclerc-Madlala 2001; Vance 1982). Farmer discusses the politics of blame as it applies to HIV/AIDS in the United States. He points that the U.S. public and medical doctors stereotype Haitians and Africans as the exotic other and associate them with bizarre sexual practices with monkeys or ritualized homosexuality and hence classify them as high-risk groups. It was implicated that Haiti was the place of origin of AIDS and hence the CDC prohibited Haitians from donating blood. As Sontag observes, "There is a link between imagining disease and imagining foreignness"—the alien, as "an exotic, often primitive place" (Sontag 1989, 48, 51).[14] This politics of blame is confirmed in China when the media blame foreigners for bringing the virus into China.

While the official media blame foreigners for bringing HIV/AIDS to China, they also stress the necessity of women's moral conduct to protect the nation's purity. In the current social and cultural change, women are targeted and held responsible for maintaining social stability and safeguarding national morality. This attitude toward women is not unique to China. Researchers have argued that women's bodies have historically been used to demarcate national boundaries (Chatterjee 1993; Mosse 1985; Stoler 1991). For instance, the rape of Chinese women by foreigners has historically been considered as the rape of the Chinese motherland (see Cook 1996). It is commonly believed that women's sexuality and bodies must be kept under strict control because women, as the bearers of national virtue and tradition, metaphorically mark the boundaries of a nation (Clarke 1999; Finnane 1996). In South Africa, the African solution to the HIV/AIDS epidemic is to support virginity testing, that is, to inspect girls periodically to see whether they are chaste (Leclerc-Madlala 2001). Virginity testing is seen as "the only way to reinstill what they view as the lost cultural values of chastity before marriage, modesty, self-respect, and pride. For them, imbuing girls with these lost values represents the surest way to repair the frayed moral fabric of society that has led to the ever increasing problems of teenage pregnancies, STIs, and HIV/AIDS" (Leclerc-Madlala 2001, 535). Leclerc-Madlala argues that virginity testing is South Africa's gendered responses to HIV/AIDS that places women at the epicenter of blame for the epidemic (Leclerc-Madlala 2001, 537). Similar gendered responses occur in China where women and girls are the target of control and the center of blame in the current HIV/AIDS epidemic.

In China, prostitutes and immoral women are blamed as the vessels and transmitters of HIV/AIDS, and teenage and adult women are placed at the center of education and training. Such an assertion of power and control over women's sexuality is meant to maintain national purity. As Jeffrey has observed, "Prostitution is a particularly rich ground for the investigation of the links between gender identity and national identity because the centrality of prostitution involves identifying correct and incorrect sexual behavior on the part of women, and distinguishing between good and bad women. Women's correct sexual behavior—usually within the bonds of marriage and family—grounds the categories of gender (what men and women should be and do)" (Jeffrey 2002, 127).

According to Foucault, "policing of sex is an important component in maintaining the unmitigated power of the central state." Women in China have historically been considered dangerous and dirty. In this time of dramatic social and cultural change, controlling women's sexuality is key to the national project. Chatterjee has stated that the nation is situated on the body of women as "chaste, dutiful, daughterly or maternal" (Parker et al. 1992, 6). Women's sexuality and bodies must be kept under strict control because women, as the bearers of national virtue and tradition, metaphorically mark the boundaries of a nation (see Clarke 1999; Cook 1996; Finnane 1996). In China, women's sexuality is strictly regulated within conjugal relationships, and women's special nurturing and reproductive role is considered crucial in transmitting virtues to the next generation and to the nation's moral well-being.

AIDS discourse is less about controlling the disease than about controlling women and promoting nationalism. I would argue that AIDS discourse has been appropriated in the service of the preexisting agendas of reasserting male hegemony and promoting a xenophobic nationalism. This is an example of Foucault's argument that modern state's discourses implant in citizens a commonsense (meaning a commonly held sense) view that facilitates the normalization and regulation of social behavior. The discourse on AIDS has become a vehicle, not for solving the AIDS problem, but for solidifying political and social control in China. Part of its genius, as indicated by these stories, is how it invites the willing complicity of women as well as men in creating national unity through the sacrifice and subjugation of Chinese women.

Although the focus of this chapter is to understand the Chinese discourse on AIDS, another issue should be addressed. What are the consequences of this discourse in stopping the spread of AIDS? Since AIDS is a relatively new phenomenon in China, we might usefully speculate about the future of AIDS in China by drawing on experiences in areas where AIDS has been longer established. Uganda in East Africa is such a place. At one time, Uganda reported 25 percent infection rates in its adult population (Epstein 2003). This has been spectacularly reversed through a campaign that approached the spread of AIDS as a sociological problem rather than a moral issue. Education campaigns urged abstinence but provided necessary education and condoms. Unfortunately, the decline in HIV/AIDS in Uganda has been recently reversed as a new campaign that assumes AIDS is primarily a moral problem has been initiated. The rhetoric of abstinence has replaced condoms, and HIV/AIDS is again on the rise in Uganda.

Our experience in Uganda and elsewhere in the world has taught us that there are social realties that must be addressed if we are to stop the spread of AIDS. It will be a tragedy for China if the health of its people is sacrificed to moralistic rhetoric designed to enhance state power rather than control a deadly epidemic.

CHAPTER 3

VILIFYING AND PROMOTING CONDOM USE IN POSTSOCIALIST CHINA

INTRODUCTION

On November 7, 1998, the first condom advertisements appeared on eighty buses in Guangzhou. These advertisements were sponsored by the Wuhan division of the UK Jissbon Global Company. The advertisement showed a messenger of love who, wearing a pair of sunglasses and a yellow robe, expressed a sweet smile to the public and promised "a love without worries or anxieties" and "a condom to ensure safety" (*quebao anquan, ziyou yitao*). This messenger of love did not live long. The company received a document about management of advertising from the Guang-dong municipal government, demanding that all condom advertisements be stopped immediately. After thirty-three days, the advertisements were stripped from the buses.

On November 29, 1999, as part of World AIDS Day education, CCTV aired a public interest advertisement that lasted forty-two seconds, featur-ing a cartoon baby in the form of a condom combating and finally driv-ing away sexually transmitted disease (STD) and AIDS viruses against a background of a newly wedded couple entering a room. The subtitle read, "Avoid unexpected pregnancy. Condom frees you from worries." A day later, this advertisement was banned by the government because it "violated the advertising law."

In 2000, a 360-square-meter-long giant advertising banner was hung up on a mansion. This time the smiling messenger survived only twenty hours before it was "killed." It was removed by order of the Bureau of Industry and Commerce.

On December 2, 2002, after the Durex Company had negotiated with various governmental departments, including the Chinese Hygiene Department, the Pharmaceutical Supervision Bureau, the Family Planning Committee for over half a year, a public-interest advertisement finally emerged on CCTV. Despite all the efforts, the Durex name never appeared on TV. The company had originally negotiated an agreement to show a public-interest advertisement on November 25 for two consecutive weeks and include the Durex brand and Qingdao Latex manufacturer. However, approaching November 21, the company received a notice from the CCTV stating that the Industrial and Commercial Bureau prohibited the display of the Durex brand on the screen. On November 26, the company received a letter from the bureau to the same effect. Instead of the ad appearing for two weeks before World AIDS Day, it was limited to one week after World AIDS Day, and the name Durex never appeared on the screen. The company thought that the World AIDS Day was a chance for Durex to enter the mass media; yet, it failed.

From November 15 through 21, 2003, the CCTV broadcast a public-interest advertisement titled, "Value Life, Prevent AIDS." Although CCTV had agreed to show the brand-name Durex on TV, the effect of the advertisement was severely mitigated by having the sponsor's name fly by so quickly that no one could catch it, thus violating the regulations requiring the sponsor's name to remain for three to five seconds. This was the second time that the Durex Company encountered setbacks on CCTV.

Durex and other condom manufacturers also attempted to advertise in media such as newspapers and magazines, but the Industrial and Commercial Bureau declared such advertisements against the law (Y. Jin 2002).

In today's world where condom marketing is perceived as a centerpiece of AIDS education and prevention, in China, as we have seen, condom advertisements in the media have been outlawed, prohibited, and severely regulated. While globally "AIDS communication programs have changed the way that condoms are perceived and promoted in many countries; this change has not yet occurred in China" (Z. Wen 2002, 14).[1] The result is disconcerting: in China, according to a 2003 national survey, 17 percent of the population had never heard of HIV/AIDS and 77 percent did not know that condom use could prevent transmission.[2] Only 18 percent to 21 percent of Beijing medical students thought condom use would protect them from HIV (Lijuan Wang 2007).[3] Only 12 percent of men from Shandong province considered condoms protective against HIV/AIDS (UNAIDS 2003). In the survey of the most recent sexual conduct, 26.1 percent reported having used condoms (14.3 percent in towns and 8.4 percent in countryside), and only 37.3 percent knew that curing other STDs could help prevent AIDS (F. Ling 2004). It is astonishing that such a high ratio of people did not know that they could use condoms to protect themselves.

In this chapter, I will investigate the conflicting meanings of, and attitudes toward, condoms by different agents driven by different interests, and the effect of the state's condom policy on the local community and the local condom market. More specifically, the state family planning administration documents routinely refer to condoms as "*biyuntao*," literally "contraception condoms," emphasizing only the contraceptive use of condoms. The alliance of local condom companies, scholars, and health realists opposed this meaning and produced "*anquantao*," literally "safe condoms," to denote condoms as a tool for safe sex. The production of these two clashing meanings of condoms is crystallized in the conflict between the state and the alliance of health activists.

This chapter will answer the following questions: What is the state's attitude toward condoms? What kind of effect does it have on the community and the local condom market? How does the alliance of local condom companies, scholars, and health realists respond to the state-disseminated hegemonic discourse about condoms? What are the underlying reasons that belie the consistent setbacks in the efforts to advertise condoms?

Through exploring the fervent condom debates by the state and the alliance of health realists and condom companies, this chapter argues that, unless the state has a proactive stance on the marketing of condom use, the empowering and persuading effect that condom marketing was supposed to have on the population cannot be reached. The impediment, in this case, the state's position and attitude toward condoms, can only thwart the progressive cause of HIV prevention.

I will first review the literature on marketing condoms and then discuss the state's attitude toward condoms in the context of the 1989 state law on condom advertisements. I will then explore the repercussions of the state's definition of condoms on local communities and the condom market. I will follow this section with an account of the attitude toward condoms by the alliance of local condom companies, scholars, and health realists. Finally, I will investigate the attitudes toward sex in the Maoist and post-Mao state and unravel the underlying reasons behind the state's taboo on condom advertisements.

MARKETING OF CONDOMS

The best means of preventing HIV is through education, which teaches the public to adjust their behaviors to reduce or to eliminate HIV exposure. Since sexual transmission accounts for most of HIV infection in the developing world and condoms are the most effective physical barrier to HIV infection, condom use and reduction of the number of partners, next to abstinence, have been the mainstays of preventing sexual transmission of HIV.

Health education involves imparting medical knowledge to the public to alter their behaviors. Earlier grassroots health educators followed Paulo Freire, David Werner, and others, and envisaged dissemination of medical knowledge as a source of empowerment (Freire 1970; Lane 1997, 166; Werner 1977). Indeed, providing the laymen with health information was believed to mitigate the level of ignorance and facilitate informed choices. Although many current agencies, such as the Ford Foundation, still operate under this mantra of health information as a form of empowerment, some researchers point out that it is difficult to measure the empowerment because of the lack of research conducted in this arena (Lane 1997, 166).

Beginning in the 1970s, there was a shift from empowerment to persuasion (Lane 1997). Market advertising techniques have been appropriated to distribute and broadcast health information to attempt to persuade people to alter their current behaviors. Since the late 1970s, a proliferation of mass media forms have been tapped into, including MTV, soap operas, and the Internet, to transform health behaviors (Birkinshaw 1989; ICAF 1989; Lane 1997, 166; Rogers et al. 1989; Singhal and Rogers 1988). The theoretical underpinning of health information social marketing is the social learning theory, which strives for behavioral change through entertainment, communication, and amusement (Bandura 1977). Entertaining programs for radio, TV, movies, or music have been employed as ideal channels throughout the world to disseminate to the public about health messages and to change behaviors (Singhal and Rogers 1999). An ideal illustration was the campaign run by Population Communication Services at John Hopkins University (Coleman 1988; Lane 1997).

In the current global HIV/AIDS pandemic, both empowerment and persuasion have been set as the strategic goal for marketing condoms. Paramount surveys have revealed that TV is the most common means through which people learn about HIV, followed by newspapers, radio programs, and journal and magazine articles. Indeed, newspapers, magazines, newsletters, television, and radio have formed a vital front line in the global struggle against AIDS. It is believed that as long as the media puts forward AIDS as a major concern in a society's consciousness, it is easier for public health professionals to disseminate preventive messages such as safe sex and condom use.

The significant role of the media has made marketing condoms a dominant approach to health education. The World Health Organization Global Program on AIDS is committed to working with mass media to enhance public knowledge about AIDS, as the organization believes that it is an effective way to help implement its global strategies against the spread of HIV/AIDS. Many countries around the world have experimented with marketing safe sex, condom use, and reduction of sexual partners. Frequent references to condoms and certain condom brands through the media have an enormous effect on people. In fact, one of the most frequently quoted reasons for condom use by the research subjects is condom advertising. People exposed to condom advertising are radically more likely to use condoms than those who have not been (Adetunji et al. 2003; Agha 1997; Ford and Wirawan 1995, 24; Messersmith et al. 2000).[4] In Tanzania, five annual surveys have revealed that radio soap operas have brought about a reduction in the number of sexual partners and increased condom adoption (Vaughan and Rogers 2000). In Nigeria, Uganda, and Zaire, studies have demonstrated that media marketing of safe sex has led to a high increase of condom use, a reduction of the number of sexual partners, and more willingness and openness to discuss safe sex (Bankole 1996; Bankole et al. 1999; Katende et al. 2000; Keller and Brown 2002).

While the literature on condom marketing has demonstrated its efficacious effect in changing people's sexual behaviors, the role of the state and churches that clashes with the marketing messages in the developing world has been either ignored or understudied. In his study of condom marketing in Mozambique, Pfeiffer delineates the contrasting messages between religious movements and condom marketers and pinpoints the role of churches in mobilizing the community to blame condom marketing for endorsing promiscuous sexuality through images and slogans and creating immoral sex and the HIV epidemic (Pfeiffer 2004). Pfeiffer's study alerts us not only to the existence of the counterdiscourse to condom marketing but also to its adverse and deleterious effect on condom marketing.

Along this analytical line of inquiry, I will use a case study from China to explore the state's role in condom marketing. I argue that unless we put the state back into the picture condom marketing will not achieve its task of HIV prevention.

STATE LAW AGAINST CONDOM MARKETING

The taboo against condom advertisements in China originated from the 1989 regulation About Prohibition of Advertisements of Sex-Life Related Products. This law stipulated that any medical equipment designed to cure sexual malfunction or to aid sexual life, although legally produced, may not be legally advertised. This law became the root of a series of setbacks to condom advertising.

State regulation was based on the state's interest in monitoring and regulating the sexual morality of its citizens. During the late 1980s, the government was concerned that society was not ready for condom advertisements because of so-called social ethics. This ethics considers condoms a product related to sex and determines that condom advertisements will encourage prostitution and promiscuity and exert a deleterious effect on children and society. Therefore, efforts to promote condoms were stifled by the state. For instance, in 2000, the police used force to close a nightclub because the club had HIV prevention flyers and free condoms issued by the Jissbon Company (T. Li 2001). The police claimed that "anywhere there are condoms is surely not a good place"; that is, existence of condoms in the nightclub indicated prostitution (Zhu 2002).

In 2001 and 2002, health professionals challenged this regulation. After long appeals, the law was finally loosened up in 2003. Although the ban was lifted to encourage public-interest condom advertisements, this change was not implemented in local areas. When I conducted fieldwork in 2007, my interviews with local government officials, managers of local TV stations, and managers of local condom companies revealed that, despite the revised law of endorsement, TV condom advertisements were still prohibited. The 1989 regulation persisted until 2007, as though the 2003 revised law never existed.

During my 2007 interviews, local TV station managers proclaimed that they had government documents mandating them not to advertise condoms. When I mentioned the 2003 revised law that lifted the ban, the answer I received was still "We have government documents that prohibit condom advertisements. After 11 p.m. though, products that cure impotence and other sexual malfunctions are advertised, but not condoms." I asked a local government official why condom advertisements were still a taboo after the law was revised, he said, "If we advertise condoms, we are issuing licenses for promiscuous sex and giving up on sexual morality. Abstinence is the best way to prevent AIDS." Other officials in the Industrial and Commercial Department contended that condom advertisements went against the socialist construction of "spiritual civilization." One official said, "China is different from the U.S. and other countries. Condom advertisements are not appropriate for China because our youth have far less sexual knowledge than their counterparts in foreign countries. The influence of condom advertisements in the media will lead them astray, away from the correct path."

This attitude is also seen in such countries as South Korea. The South Korean government strove to revive Korean values of purity and morality and deemed condom and contraception education for the young as "an uncritical adoption of western-style sex promotion," hence "culturally inappropriate for Asian youths" (Cheng 2005).

Religious groups worldwide have voiced similar anxieties about condoms (Pfeiffer 2004; Smyth 1998). In Ireland, for instance, the Catholic Church believes that the wide availability of condoms serves to heighten the problem of HIV/AIDS (Smyth 1998). As a result, the Catholic teaching that sex must be confined to marriage has permeated the government's response to AIDS. Indeed, this message has been reinforced not only in the government's advertising campaigns but also in their educational and informative materials (Smyth 1998).

My interviews in Dalian have revealed that the revised law was not operationalized in local communities because of the state's concern that condom promotion can increase immoral sexual activity and promiscuity. I argue that this concern arises from the government's interest in disciplining the sexual morality of its citizens, thus defining condoms as a contraceptive tool that should be used only within the bounds of marriage rather than to prevent venereal diseases. Indeed, despite the state's concern, studies conducted worldwide have consistently showed that encouraging condom use does not increase sexual activity. It simply makes sexual activity safer (Guttmacher 1997; Sellers et al. 1994).[5]

EFFECT ON LOCAL COMMUNITIES AND UNIVERSITIES

The state's attitude toward and definition of condoms as a contraceptive tool, to an extent, has penetrated local communities, including universities. During my research, when I mentioned the free condom distribution program by local nongovernmental organizations (NGOs) to local people, they appeared alarmed, shocked, and confused. They responded, "Handing free condoms to everyone—how can our country allow them to do that? Doesn't that encourage and promote promiscuous sex? Giving people free condoms—doesn't that endorse illicit and random sex? Won't the society be in disorder with such immoral sex?"

Local communities and many key universities in the country are in line with the state in defining condoms as a contraceptive method employed within marriage. In 2004, despite the call from the Beijing and Hubei Hygiene departments to prevent venereal diseases, no universities allowed them to install condom vending machines or distribute free condoms on campus (F. Jing 2004; W. Zhang 2006). During my research in two local universities, one of which was a medical university, local professors were careful not to "corrupt" their students with discourses on sexuality. When I showed them my survey on HIV knowledge, they crossed out all the questions that contained the word "sex" before distributing the surveys to the students. As a result, half the survey was deleted. When I asked for the reason, the professors looked at me as if I came from another planet: "We can't expose the students to these sexual ideas. They are too young to know this stuff. Knowledge about sex can only arouse their curiosity and encourage them to try it out. It's too dangerous for the students to know about this stuff."

Some professors at the local universities rejected proposals from local NGOs to educate the students about condoms. They contended that the school was different from society and that condom distribution or education on campus was inappropriate because few students engage in sexual activities; hence, the program would only initiate and encourage their sexual practices. Professor Wang Wei from National Executive College spent ten years on the book *Sex Ethics*, arguing that despite the importance of condoms, they are, after all, special merchandise that should be available for purchase but not seen everywhere in the society (Rong 1999). Professors such as Wang Wei believed that using condoms to prevent AIDS had turned what should be a moral issue into a technical issue. They argued that sexual morality, rather than condom use, should be emphasized on TV.

At times, these university professors' stance was co-opted and reinforced by foreign Christian groups who preached abstinence only in Chinese universities. For instance, three American "sex-education experts" arrived in Beijing on October 10, 2004, and spent a week delivering speeches on abstinence only at middle schools, colleges, universities, and other community locations. They warned China not to relive the errors that the United States had committed for thirty years, that is, emphasizing condom use and not emphasizing abstinence as the sole safe choice. They stressed that "sex is only beautiful when it happens within a marriage" and that condom use is only applicable to prostitution (Li et al. 2004).

The emphasis on the deleterious effect of condom use also appeared in some Chinese media: Condoms could harm a woman's health because women would be deprived of the sperm that could reduce vaginal infections, fight ovarian cancer, boost female hormone production, and produce robust breasts and tender skin (An 2006).

The state's attitude toward condom use and definition of condoms as a contraceptive tool only appropriate within marriage has influenced local communities and some university professors. Indeed, it was stipulated as early as 1990 in the Basic Requirements for Health Education formulated by the Chinese Education Ministry that sexual morality and self-discipline should be taught to prevent HIV/AIDS and STDs (Chiang 2004). Moreover, the National Education Department deems any sexual conduct of students as severe transgressive behaviors, and students who engage in sexual behaviors are either expelled from universities or detained at school (W. Zhang 2006). As we have seen, university professors and local communities speak the state language in associating condom promotion with sexual promiscuity and prostitution and defining condoms as a contraceptive method legitimate only within the bounds of marriage.

IMPACT ON THE CONDOM MARKET

The state's attitude toward condoms and the taboo on condom advertisements have greatly affected the society. During my research, when I asked the clients what kind and what brand of condoms they purchased, clients looked confused and baffled. They asked, "What kinds of condoms are there? I don't know anything about the kinds or brands of condoms. If you know about this, could you please tell me what kind and what brand of condoms are better?" Of course, I did not expect this answer at the beginning, but as my research continued, I realized that people were ignorant about condoms because it was a taboo topic in the media and in people's conversations.

The taboo against condom advertisements has left people uninformed about condoms and generated embarrassment in purchasing condoms; even young female customers are equated with sex workers. Clients in my study said that when they purchased condoms they always dropped their heads, grabbed the condoms, and escaped swiftly out of the store as if they were prisoners at large. When I asked how they chose which condoms to buy, they told me that it was too embarrassing to have others see them purchasing condoms. They had no idea which condoms they bought or the price because they either took a random pack from the counter or asked the counter staff to hand them a pack and left immediately after payment. As for women, during my research, some young local women told me that although they were married because they looked young and unmarried, they encountered stigmatizing comments when they purchased condoms. A twenty-eight-year-old woman told me,

> Although I am married, people always say that I look like I am 21. One day after work, I had time to stroll along the street and happened to step into a drug store. Dozens of colorful packages of condoms lying under the counter caught my eyes and aroused my curiosity. I thought I should get one and try it out. After studying them for a while, I still had no idea which one I should buy. So I turned to the shopping assistant, asking: "Could you please recommend one with good quality?" She looked me up and down, and then sneered at me, saying, "You don't know? You should go ask your clients!" I was dumbfounded. I was so angry with her words that I stood there and could not say a word. Tears rolled down my eyes. I left the store, crying the whole way home and swore that I would never buy condoms again.

In Dalian, couples of a reproduction age must bring their marriage and reproduction certificate and temporary resident card to get a one- to three-month's supply of free contraceptives from the city family planning office (Lao 2005).[6] The state's strict stipulations on condoms have reinforced local communities' attitudes toward condoms. As condoms are considered appropriate only for married couples, any nonmarried consumers are deemed sexually promiscuous.

Local communities elsewhere in the country have also spoken the state language. For instance, in 1998 when condom advertisements first appeared on buses in Guangzhou, many citizens called 110 (the emergency number) or sued the condom company, claiming that their advertisements corrupted the souls of the youth and harmed the morality of society (Rong 1999). In Suzhou, on World AIDS Day in 2001, a local woman called 110 and reported to the police that she saw people distributing free condoms on the street. In 2006, citizens of Chongqing criticized the pilot program of 100 percent condom use in adult entertainment places, contending that the program approved of and encouraged extramarital affairs and prostitution (Yin 2006).

The taboo against condom advertisements has not only penetrated the society with the state's hegemonic view about condoms but also led to a lack of competition in the condom market, which fueled fake condom brands. Although family planning offices distribute free condoms, local people told me that those condoms were of such low quality that they had to revert to the stores to purchase condoms of better quality, styles, and colors (T. Dong 1999).

I visited several local adult health-product shops and interviewed the owners about the condoms. An owner of an adult health product shop told me that she used to be the leader of the city's family planning office during the Maoist era. She told me many stories about condoms during that time.

> People at that time came to me complaining that condoms did not work because their wives continued having babies. So I asked the guys how they had used the condoms. The men put the condoms on their thumbs and said that was how they had used them, just the way it was demonstrated to them when condoms were distributed. Others used condoms as balloons rather than as a contraceptive tool . . . Tons of unused condoms were stored during the 1950s and 1960s. These include condoms from the family planning office, expired and saved from free distribution. After they expired, officials in the office privately sold them cheaply to individual vendors. Vendors then recycled and repackaged them in brand new and sexy boxes, and bribed managers of supermarkets and pharmacies to have the products sold in these stores, at a high price. They looked new, but indeed, they were not new at all. They were either expired condoms from the family planning office, or the old items stored for decades.

As stated by the store owner—a leader of the family planning office during the communist age—people were ignorant of how to use condoms and the cadre's demonstrations were vague and opaque. Apparently, even back then, the condom was an embarrassing topic because of its relationship with sex. This points to a historical continuity until the post-Mao era, which I will elaborate on later in the chapter.

The store owner also referred to the proliferation of expired, fake brands in the condom market, some of which were repackaged old products stored for decades. This story was confirmed by other store owners of adult health shops and drug stores. I talked to the manager of a local condom manufacturer about this story, and he said it was illegal for the family planning office to sell the expired products saved from free distribution to individual vendors; however, those products were indeed on the market. Some condoms were not even repackaged. As I window-shopped, it was easy to spot un-repackaged, expired condoms for sale. "Not for sale" (*fei mai pin*) would be printed on the old packages. There was no information about the expiration date or the manufacturer. It was obvious that they came from the family planning office. Other expired condoms were repackaged in boxes with images of naked white women.

Store owners told me that 70 percent to 80 percent of these condoms broke. They were very easily torn open due to the expiration and poor durability. A box of ten condoms sold for 6 yuan. According to the owners, people favored these condoms because of their low price. As discussed earlier, some of my research subjects indicated that they were too shy to inquire about condom quality. Rather, they asked for a recommendation, purchased them, and left immediately. In recommending condoms, would the shopping assistant consider the well-being of the customers? My research shows that the managers and sales persons received bribes from the suppliers. They would only push the products of those from whom they had received the biggest bribes and exhibit their products at the most highlighted place on the sales counter—at the center.

The taboo against condom advertisements and the lack of competition have resulted in a market with poor-quality, expired condoms.[7] In Wuhan, for instance, it was reported in 2001 that 70 percent of the abortions were caused by faulty condoms (W. Lin 1998). In the same year, a national survey conducted by the National Quality Control Bureau reported that a plethora of packaging businesses purchased obsolete or discarded condoms, wrapped them in colorful and pornographic packages, printed some international famous brands on them, and sold them in the market. This led to an inflation of brand names from ten at the beginning of the reform to more than one thousand, including fake inland or foreign brands that packaged the same types of condoms. In 1998, more than 5 million fake condoms were discovered, confiscated, and burned in Futian and Shenzhen (W. Lin 1998). The market leaves consumers confused, without sufficient means to choose condoms intelligently.

ALLIANCE OF CONDOM COMPANIES AND HEALTH PROFESSIONALS

In opposition to the state's definition of, and attitudes toward condoms, the alliance of condom companies, scholars, and health professionals defined condoms as a disease-control tool, a personal hygiene product, and a health product. As a result, they desexualized condoms and pushed for endorsement of condom advertisements.

The 1989 law and the series of setbacks to condom advertising ushered in a lively debate initiated by health professionals, scholars, and condom companies (W. Lin 1998). Li Jihong, the deputy secretary of the Sexology Committee, considered it crucial to encourage condom companies to support the dissemination of HIV information by endorsing condom brands to appear in public-interest advertisements (Y. Jin 2002). The manager of the Jissbon Company stated that condom advertisements would be beneficial in establishing brand names, increasing sales of condoms, and helping to push "progressive ideas" (Yi 2001). Xuehai Wang, the president of Wuhan Jissbon Company, also expressed his determination to continue the advertising efforts so that in the future, buying condoms would be no different from buying shampoo; that is, there would be no embarrassment or humiliation involved in condom purchases (J. Chen 2002).

The alliance showed the urgency of the matter from the three aspects. First, one Chinese male uses no more than four condoms each year, and unprotected sex is still an important venue for HIV transmission (Yi 2001). Second, premarital sex has become prevalent in society, even in schools.[8] Third, China is facing a HIV epidemic, and only a few people know that condoms are an effective tool against STDs and HIV/AIDS (J. Li 2000; T. Li 2001; W. Lin 1998).

Condom advertisements could improve this pressing situation, as they contended. Health professionals and scholars believed that condom companies' financial support was indispensable in disseminating prevention education (Yi 2001). Indeed, many companies had already issued prevention booklets and provided free condoms on World AIDS Day. Because condom companies lacked media support, they had to resort to the street to issue HIV flyers and free condoms or install condom vending machines in communities and campuses (Liang 2003; J. Zhao 2004).

The alliance of condom advocates considered the ban on condom advertising adverse to social progress and civilization and criticized the state policy as an "ostrich policy," which avoided sex issues and demanded purity (S. Li 2000). They compared their cause to cutting queues in the Qing dynasty, ending footbinding, and allowing girls to go to universities. To them, the debate was about the choice between life and ethics, but different from the other causes, the cost of this debate was life.[9]

The alliance was committed to having the 1989 law revised. In March 2002, Li Honggui, vice president of the Chinese Population Association, forwarded a plea to the National Congress to lift the ban on public-interest condom advertisements under the condition that the ads were under government supervision. Li contended that defining condoms as sex equipment and banning condom advertisements prevented the consumers from obtaining information from normal channels, stymied the enhancement of condom quality, and hindered the establishment of superior condom brands (Anon. 2002). The plea was signed and supported by more than one hundred representatives of medical and other professions. In June, the Industrial and Commercial Bureau replied to the plea, agreeing to lift the ban and allowing condom advertisements "under special conditions and with limitations" (H. Fan 2002).

After long appeals for condom advertisements by health professionals and scholars, June 2003 witnessed a change in regulations, whereby *limited* condom advertising was granted. In July 2004, About Condom Use for HIV/AIDS Prevention regulation was promulgated, which encouraged public-interest condom advertisements for disease prevention (Z. Wang 2003).[10]

Although it was a victory for the alliance of condom companies, health professionals, and scholars to revamp the 1989 state law, my research in 2007 revealed that the taboo against condom advertisements persisted in local areas. Despite allowing condom advertisements "within limitations," social stigma continued, associating condoms with prostitution (H. Yu 2002).

CONCLUSION: BEHIND THE TABOO
AGAINST CONDOM ADVERTISEMENTS

The taboo against condom advertisements spanned more than fifteen years and only loosened up in 2004, after years of debates and appeals from hundreds of health professionals, scholars, and condom companies.

The condom debate crystallizes two opposing definitions of condoms, driven by multiple-positioned interests. The state defines condoms as a contraceptive method associated with sex within marriage and believes that condom advertisements will encourage prostitution and promiscuity. The alliance of condom companies, health professionals, and scholars, however, defines condoms as a personal hygiene item, a health product, or a disease-control tool and desexualizes condoms.

Medical professionals and scholars believed that condom companies could provide necessary financial support for HIV information and more prevalent condom use. Guided by their interest in ameliorating the pressing HIV issue in China, they appealed for an overhaul of the state law and supported condom advertisements as an emblematic sign of social progress and a civilized nation. Driven by commercial interests, condom companies allied with medical professionals and scholars in propelling the cause.

The state's stance on condoms has affected and, to an extent, even penetrated local communities and universities and fueled the sale of unqualified, expired condoms in the condom market. Taboos on condom advertisements and social stigma associated with condoms have contributed to popular ignorance about condoms and the prevalence of unsafe sex.

What about the interests of the state? What has guided the state to prohibit condom advertisements? What are the reasons behind the taboo?

Chinese sociologist Li Yinhe explains that because the state construes sex as essentially bad, they are concerned that people might have fantasies about sex when watching condom advertisements. Such fantasies about sex are deemed criminal, debased, and indecent (Yinhe Li 2002). Li contends that before the Song dynasty sex was regarded as natural and healthy, as it facilitated the coalescence of yin and yang as the principle of life. Sexual desire was only demonized and criticized as against nature after the Song dynasty. Li points out that the dearth of sexual discourse during the Maoist era still lingers in the present society. Once we change our ideas about sex, she argues, it will be easy to deal with condoms (Yinhe Li 2002).

I agree that behind the regulation is the state's worry about citizens' sexual fantasies. But what accounts for such a worry? Why does the state consider sex so dangerous? Why does the state monitor and regulate citizens' sexual morality?

To understand the post-Mao state's attitude toward sex, we have to recognize that the Maoist legacy still persists. During the Maoist era, sexual desire was demonized and criticized as capitalist love, which degraded love to animal instinct and reduced love to prostitution. In 1954, the journal *Women of China* published an article on Lenin's speech on sexual ethics (Q. Ma 1954). The article quoted Lenin's talk, which derided and criticized "a glass of water-ism"—treating sex like drinking a glass of water. According to Lenin, it is dangerous to favor sex because sex is a lowly, decadent, and corrupt thing. Sex is so pernicious that it robs people of happiness and strength.

Revolution, Lenin says, requires strength and aggregate energy, not intoxication and stimulation from sex or alcohol. It does not tolerate degenerate lewdness. Sex is incompatible with revolution, so there is no future for the people who favor sex. Lenin states that he would never trust those who enchant women or fall in love numerous times. He calls on people not to weaken, waste, or destroy their strength through sex. Rather, energy should be reserved for revolution, and sexual desire should be sublimated for the revolution.

Lenin was cited during the Maoist era to emphasize that self-restraint and self-discipline should be exercised in sex because family is the basic cell of society and the state, and marriage and love are not personal but social in nature as they produce new lives and new responsibilities. It is said that marriage is such a serious cause that people who do not marry or produce children are not responsible for the future of human kind, and it is important to choose a spouse who is patriotic and supports communist construction. Careless divorce and remarriage are not forgiven by the society because we need to guard social morality. Healthy love and a happy family can stimulate creative work for the communist state (Q. Ma 1954).

The Maoist era exhibited a single, monolithic state voice that underscored the absolute conflict between sex and state. That is, sex saps, weakens, and debilitates people's energy that should be devoted to the state, and hence, sexual desires should be sublimated to construct socialism and contribute to the state. Sex is only legitimate when producing the next generation for society.

Limiting sex to procreation resonates with Foucault's concept of "alliance" rather than "sexuality." In *The History of Sexuality*, Foucault depicts two ways of governing—alliance and sexuality. Alliance is "a system of marriage, of fixation and development of kinship ties, of transmission of names and possessions" (Foucault 1978). In the system of alliance, reproduction is the ultimate goal. The transformation of alliance to sexuality began in the last third of the eighteenth century in Europe when modern science recognized a fundamental, biological difference between the male and female sexes (Laqueur 1990). This naturalized and essentialized view of sexual difference with a biological basis led to the state's concern with "the sensations of the body, the quality of pleasures, the nature of impressions" in the system of "sexuality" and enabled the state to penetrate families and private bodies "in an increasingly detailed way" (Foucault 1978, 106–7). Rather than centering on reproduction, "sexuality" focused on "the exploitation of the body" that "produces and consumes" (Foucault 1978, 106–7). According to Foucault, in modern Europe, "sexuality" was superimposed on the old system of alliance and reduced its importance.

In late-imperial China, gender was anchored on social roles, and women were defined by their roles as mother, daughter, and wife. Reproducing the lineage was the meaning and purpose of sexuality. The May Fourth Movement initiated a biological definition of male and female in 1919 (T. Gao 1919). The biological and unitary category of women— *nuxing* (female sex)—was created during this time in place of *nu* (daughters), *fu* (married wives), and *mu* (mothers; Barlow 1994). For the first time in Chinese history, there was a word meaning biological woman (M. Yang 1999a). The Maoist state suppressed and obliterated this biological difference between male and female and created an ostensibly androgynous gender model where men and women wore unisex clothing and femininity was rejected as bourgeois.

The Maoist era's emphasis on alliance restricted sex for reproduction as it benefited society and helped build the state. Sexual pleasure was castigated as a feature of the decadent and degenerate capitalist lifestyle. Hence, policing and disciplining sexual activity that was considered not contributing to state building prevailed.

While the Maoist era reviled sex as dangerous and antithetical to state construction, the post-Mao era witnessed a cacophony of conflict between the state and alliance of health professionals and condom companies. The post-Mao state inherited and perpetuated the Maoist state's stance that sex is only legitimate within marriage. Cohabitation is rejected and prostitution is outlawed. The government stages annual "anti-vice" campaigns aimed at eradicating the "ugly phenomenon" of prostitution. Western sexual liberation is debunked, and books that are deemed to reflect this theme of sexual liberation are banned, such as *Shanghai Baobei* (*Shanghai Baby*) and *Fei Du* (*A Decadent Capital*). The ban on condom advertisements is one of the myriad examples of the state's denial of sex outside of marriage.

However, whereas the Maoist state rejected sexual pleasure, the post-Mao state recognizes the importance of sexual pleasure *within* marriage as it maintains marital harmony and thwarts extramarital affairs (Sigley 2001). A "socialist sexual morality" is emphasized to ensure a harmonious conjugal family as it is critical to secure social stability and state control.

What also differentiates the post-Mao state from the Maoist state is the cacophony from the alliance. As the condom debate reveals, while many of the local communities are influenced by the state and speak the state language in insisting on abstinence and sexual morality as the ultimate antidote for disease prevention, the alliance of condom companies and health professionals recognize the current prevalence of unsafe sex and employ the discourse of "social progress" and "civilization" to promote condom advertisements. Some even rebel against the state's "repression of sex" by subscribing to the concept of "natural" sexual desire.

In a previous article, I discussed the ways in which many businessmen, entrepreneurs, and nouveaux riches romanticize and desire a natural body filled with natural instinct, free from the unnatural socialist system (T. Zheng 2006). They engage in sex consumption to convey their disbelief, criticism, and defiance of the socialist system. According to them, socialist ideals of self-sacrifice and self-discipline have lost currency among the Chinese people because socialism and the Chinese Communist Party are no longer trusted. Caught in this moral vacuum, they turn to the hedonistic pursuit of sexual pleasure and material wealth to compensate for their emptiness and meaninglessness. Sex consumption is regarded as resistance to the artificial shackles placed on human sexuality by an unnatural social system. For many of these modern hedonists, socialism is an oppressive regulatory regime that stifles "human nature" and goes against the "natural way." Here, nature is defined according to their (and most mainland Chinese people's) perception of the West, in particular, the United States, as a land of free and open sex.

Such illicit sex and free pursuit of sexual pleasure represent these people's resistance, disbelief, and defiance of the socialist state. Posing a direct challenge and menace to the state, illicit sexual pleasures are the main target of state control. As Foucault contends, "policing of sex is an important component in maintaining the unmitigated power of the central state" (Foucault 1978, 24–26). If sex can be regulated and maintained within the confines of marriage, then social stability and state control is ensured. Thus, it is essential that "the state knew what was happening with its citizens' sex, and the use they made of it, but also that each individual be capable of controlling the use he made of it" (Foucault 1978, 26). Reshaping and disciplining sex is hence central to the state's desire to maintain and secure the social fabric.

The post-Mao state has exercised many strategies to negate, to silence, and to police nonmarital sex. These strategies include prohibiting condom advertisements, banning books with sexual content, and cracking down on prostitution with anti-vice campaigns. The state also propagates sexual morality, satisfaction of sexual desire within marriage to secure its stability (Sigley 2001), and laws and regulations stipulating permitted and forbidden sexual behaviors. As Foucault observes, "What is peculiar to modern societies, in fact, is not that they consigned sex to a shadow existence, but that they dedicated themselves to speaking of it *ad infinitum*, while exploiting it as *the* secret" (Foucault 1978, 35). Like Freud, the Chinese state believes that the price of civilization is repression. The state's obsession with sex and with controlling it is seen as necessary to achieve what the state envisions as a harmonious society, in which citizens are obedient and the state is in control. The state's prohibition against condom advertisements is one of the interventions that are aimed at concealing and suppressing nonmarital sex to ensure the stability and harmony of marriage and family on which the state establishes and maintains its power and control.

CHAPTER 4

PERCEPTIONS TOWARD CONDOM USE AMONG MALE CLIENTS OF DALIAN HOSTESSES

INTRODUCTION

THE MEANINGS OF MASCULINITY EVOLVED THROUGHOUT CHINESE history. Before the May Fourth Movement in 1919, courtesan house was a site that produced an elite masculinity of self-control and cool demeanor. Elite masculinities had to be validated by the courtesans, the arbiters of their masculinity, as worldly, urbane, knowledgeable, sophisticated, and refined (Henriot 2001; Van Gulik 2003).[1]

These values were so important that there were guidebooks instructing men in appropriate conduct. A customer had to learn the aesthetics and etiquette of frequenting courtesans to obtain respect from other men, avoid ridicule from courtesans, and demonstrate his sophistication. With inside knowledge of the brothel, he would know how to deal appropriately with courtesans and madams. For instance, he would be able to manage the tea ceremony (the best time to build an intimate connection) skillfully and with face. He would spend time establishing a relationship with the house and a courtesan, fulfilling his obligations such as hosting banquets or gambling parties. He could demonstrate his connoisseurship through his appreciation of the artifacts decorating the courtesan's body and by describing them eloquently. He would know how to judge whether he was treated by the courtesans with due deference. He would be careful to exhibit not only wealth but also good taste in dress. He would go to the courtesan house in the company of powerful male companions, so that the courtesans would not dare to play tricks on him. He would pay attention to details such as helping her into her coat or seeing her off. A customer who failed in these manners lowered his own status and risked exposure in front of his fellow customers as a "country bumpkin" (Hershatter 1997, 69–102).

Concerns about masculine identity at this secure time of "culturalism" (Fitzgerald 1996) have to do with social class and have no reference point outside of China. With the Western intrusion into China, Chinese male insecurity was linked to the perceived decline of China and contributed to the growth of Chinese nationalism. Elite masculinity was attacked because it was identified with the elite cultural tradition (Larson 2002). Nationalism produced a new model of masculinity. The powerless and impotent son had been replaced by a rebellious and sexually powerful male. For the first time in Chinese history, the sexual prowess of Chinese men was not measured internally as a means to establish social class but came to be measured against the outside predators whose military prowess identified them as more sexually potent.

Later on, the Maoist state, with its emphasis on gender equality, attempted to control men's sexuality by suppressing female sexuality (T. Zheng 2009).[2] As I revealed in my previous book *Red Lights: The Lives of Sex Workers in Postsocialist China* (T. Zheng 2009), in the 1980s, the suffering of women was seen as a way to ensure men's success and redeem the nation's disgrace (Brownell 2000).[3] In the 1990s, masculinity and marital stability were seen as dependent on women's enjoyment of sex. This radical notion that women should enjoy sex was not out of a concern for the happiness of women, but rather reflected the new competitive capitalist economic model where men proved themselves through entrepreneurial activity. Men were judged not by birth status or even education but by their competitive abilities. The effect of this change on the relationship between men and women was profound. Women became a testing ground for male entrepreneurial ability. In this competitive world, men's skill in charming women and keeping them under control came to define their success. Hence, in the new fluid urban entrepreneurial environment, men attempted to rebel against the Maoist control of their sexuality and resurrect their lost masculinity by emulating the economically successful Taiwanese businessmen in the consumption of women.[4] Their subjugation of women represented the recovery of their manhood in post-Mao China.[5]

This new meaning of masculinity reflects what researchers such as Micollier and Hyde have deemed as the conflicting sexual moralities that "simultaneously challenge and embrace Confucian notions of Chinese sexual behavior and Maoist notions of restraint." Hyde names these competing sexual moralities as "parochial Maoist morality" and "a liberal market morality" (Hyde 2007, 124; Micollier 2005a). Researchers have pointed out that globalization, the market economy, the one-child policy, and China's transformation to a consumer society have separated sex from reproduction, detached sex from love, and catapulted the sexual revolution and promoted a liberal model of sex for leisure (Farquhar 2002; Farrer 2002; Hyde 2007; McMillan 2006; Micollier 2005a; S. Pan 2006; Rofel 2007). This morality, according to Hyde and other researchers, is epitomized by "prostitution, multiple lovers, and unprotected sex" and is controlled and regulated by the "footprints" of the previous Maoist regime, or the so-called parochial Maoist morality (Farquhar and Zhang 2005; Hyde 2007, 170–71).

This liberal market morality is epitomized by leisure sexual activities in recreational spaces such as karaoke bars and sauna bars. The state, however, still supports the parochial Maoist morality, constraining and legitimizing sex only within a conjugal, heterosexual family, and deeming leisure sexual activities as illegal (Gil et al. 1994; Jeffreys 2004, 2006; Ruan 1991; Sigley 2001; Zhong 2000). As a result, hostesses and male clients are under constant state surveillance, subject to random harassment and arrest by the police (T. Zheng 2009).

AIDS researchers have argued that "the pandemic cannot be understood without reference to the capture of the state in many areas of the world by conservative forces seeking to control sexuality and pleasure" (Bond et al. 1997; Feldman 1994; Schoepf 2001, 352; Seidel 1993). In China, what is the relationship between the state control of sexuality and pleasure and male sex consumers' condom use? How are condoms connected with the state's repression of sexuality? How does this connection affect clients' decisions about condom use? How much HIV knowledge do clients have, and how do they perceive condoms and sexually transmitted diseases (STDs)?

In this chapter, I will first assess the clients' medical knowledge about the risks of unprotected sex and the benefits of condom use and then explore clients' perceptions of condoms in their social and cultural context. The data are drawn from questionnaire surveys, participant observation in a condom manufacturing company, and interviews of clients, hostesses, and owners of adult health shops.

MEDICAL KNOWLEDGE

Do clients know about the risks of unprotected sex and the benefits of condom use? I conducted both questionnaires and interviews to test the clients' medical knowledge about HIV/AIDS and condom use. The questionnaire consisted of twenty-five questions testing their knowledge of HIV/AIDS, ten questions on the use of condoms, and twenty-one questions on their attitudes toward HIV/AIDS patients. If we control for age, male clients (63.15 percent) scored nearly as well as male (66.95 percent) and female (65.85 percent) professors from a medical university in Dalian. For the question "Can condoms effectively reduce HIV transmission?" the clients answered correctly 100 percent of the time, compared with only 48 percent by male hotel employees and 57 percent by female hotel employees.

One thing that needs to be noted is that although 100 percent of the clients are aware that condoms can prevent HIV and STD transmission, they lack some detailed knowledge about how HIV is transmitted. In my interviews with the clients, no one knew that HIV could be transmitted through oral sex. Nor did anyone know that that HIV could be latent in the body for years before manifesting any symptoms. Some even thought that sharing toothbrushes could transmit the virus. All interviewees believed that in China HIV was mainly transmitted through blood transfusion, commercial selling of blood, or intravenous drug use. Transmission of HIV through sexual intercourse was possible, according to them, but very unlikely in China.

Nonetheless, clients still exhibited the highest level of knowledge of HIV and condoms of all the groups. Given this, one would expect the clients to use condoms consistently to protect themselves. However, many clients wrote directly on the survey that they did not use condoms because they did not feel comfortable. None of the clients responded to the question about reasons for using a particular brand, indicating that it might be uncommon for them to either use or buy condoms. In my subsequent interviews with the clients, I learned that they never used condoms with their wives because their wives had intrauterine devices (IUDs). Client Huang said, "I only used condoms during the one month before my wife gave birth. Ever since she gave birth and had an IUD, I quit using condoms." This information coincides with studies conducted in Nigeria where condom use with wives is considered unnecessary because of trust and intimacy (Messersmith et al. 2000). However, in contrast with Nigeria where men report consistent condom use with sex workers,[6] clients in my study related inconsistent use of condoms with hostesses. Client Huang said that he did not use condoms with hostesses during his business trips with his superior. When his superior suggested buying hostesses for the night, he followed his lead and engaged in illicit sex with the hostess his superior chose for him. Conducting illicit behaviors together and sharing the same secrets formed a bond between him and his superior and cemented their relationship. He said the first time when he was having intercourse with the hostess, he was so nervous that he finished it in a couple of minutes. Condoms never crossed his mind because he had never used them before. When I asked whether he feared venereal diseases and HIV, he responded that all his friends had contracted venereal diseases, but since they are all curable, it does not constitute a real worry. In his words, "my friends just went to the hospital and got a shot each time when they had a sexually transmitted disease, and they were healthy and fine afterwards."

CULTURAL FACTORS

As research in many African countries has indicated, a gap remains between knowledge about condoms and use (Amadora-Nolasco et al. 2001; Messersmith et al. 2000; Wulfert and Wan 1995). Research findings have repeatedly shown that correct answers about HIV were irrelevant to consistent condom use, pointing to the complexity of risky behaviors. As Schoepf notes, it is the local sociocultural processes that create risk of infection (Feldman 1994). Research has shown that interpretations of cultural meanings are central and crucial to understanding the reasons for this gap and sexual transmission of HIV (see Micollier 2004a; Parker 2001).

Next, I argue that cultural factors determine clients' nonuse of condoms. These cultural factors include the perception that condoms represent unnatural state control over natural sexual expression, the perception that contraception is women's responsibility, and the perception by the peer group that nonuse of condoms is brave and valiant (*meng*).

CONDOMS AND STATE REPRESSION OF SEXUALITY

In China, condoms were closely associated with a planned economy and population control. During the planned economy, Dalian Latex Factory was established in 1958 as one of the seven factories appointed by the government to produce condoms.[7] The state purchased condoms from these factories and distributed them to different areas free of charge. The brand name of the first condom made in the Dalian Rubber Factory was "family planning" (*jisheng*), produced in 1975 and trademark registered in 1979. Instead of being sold in the market, these condoms were distributed free of charge to units and community committees of counties, districts, and cities.

During the first stage of condom production, in the popular mind, condoms were deeply infused with the power of the state and the state control of sexuality. The government chose "family planning" as the brand of condoms and called condoms "contraceptive condoms." Both the brand and the name of condoms denoted the purpose of condoms as contraceptives for population control. Condoms and other contraceptives have been enforced by the state in accordance with the policies of population control since the 1980s. Condoms were referred to as both "safety condoms" and "contraceptive condoms" during the late 1990s. Despite these new changes, today drug stores, supermarkets, and convenience stores place condoms under "family planning products," indicating that condoms are for contraceptive purposes.

As Ann Anagnost (1995) suggests, Foucault's concern with sexuality in Western societies as pivotal focus of political, economic, and social interventions can be replaced in China with population and reproduction (see also Sigley 2001, 132). As illustrated, condoms have been imprinted with state power, the power to control sexuality and population. Condoms were defined as contraceptives and enforced as a contraceptive among married couples to reduce the birthrate. Hence, historically, control of condoms is realized through controlling its name, its brand, its appropriate users, its purposes, and its distributions. Controlling condoms associates condoms inextricably with state repression and control of sexuality.

Research elsewhere has also reported that coercive population control imposed by the government provides a moral context in which condoms are branded with population control and malicious authority (Kaler 2004). For instance, across Africa, family planning programs have been viewed as a vicious and brutal way to wipe out the African population. In countries such as Nigeria, Tanzania, and Malawi, family planning is construed as a postcolonial conspiracy of Western countries and the white population to control the African population. In Tanzania, it was believed that white people put viruses in the condoms intended to harm them (Plummer et al. 2006). In Malawi, people believed that condoms were sent from America to reduce the Malawian population (Kaler 2004).[8] For people such as Cambodian refugees in Thai refugee camps; women in Chiapas, Mexico;

women in Indonesian-occupied East Timor; and women on aboriginal reserves in northern Canada, family planning methods were perceived as the government's hidden agenda to annihilate the native population. The same rumor was reported around the globe in countries such as Niger, Kenya, Cameroon, and Bolivia, inducing a general suspicion and fear of condoms. Likewise, in China, condoms are entangled with "the symbolic nexus in which they are fused with disease, population control, and the malevolence of authority" (Kaler 2004, 114).

ANATHEMA TO CONDOM USE

As I elaborated in the methods section in the introductory chapter, I encountered immense resistance to the topic of condom use and HIV during my research. When I served as an AIDS volunteer worker at the local Centers for Disease Control and Prevention, the director warned me not to distribute free condoms and be extra cautious when disseminating AIDS booklets. As she explained,

> I cannot give you the condoms because if you go there on your own and hand out the condoms to the manager, it may cause trouble. They may call the police and have you arrested. So don't distribute the condoms, and be extremely careful when you hand out the booklets. If they don't want it, don't force it. It's totally up to them whether they want it or not. You should not force anything on them. Nowadays people don't have the consciousness to learn about AIDS. People are strongly resistant and antagonistic towards it [*dichu qingxu*]. It will infuriate people if you give them this little book. They are very averse to this kind of information.

This kind of resistance was obvious throughout my research. My interviewees resisted condom use and were uncomfortable with questions about condom use. During my research, clients were antagonistic to my questions about HIV/AIDS and condom use. When I talked to an interviewee about HIV/AIDS, he brushed it away saying that AIDS was mostly transmitted through blood transfusion in China. When I asked about condom use, he said, "I don't care—whether to use it or not to use it" and ended the conversation. I approached a friend of mine, the general manager of an import and export company, and asked whether I could hand out a questionnaire to him and his employees about HIV/AIDS knowledge and condom use. He said, "It's impossible to do this here. Chinese people think differently from the Americans. My employees would be wondering what I'm doing. They're all very repulsed [*fangan*] by this kind of thing. I'm sorry that this cannot be done here. You cannot force me." When I pressed him on why people are "averse to this kind of thing," he did not explain but insisted that this could not be done here because this was China.

Why is it so difficult to have a conversation with clients about condom use? Why did these men have a lot of interest in talking about hostesses but not condom use? Why did they feel repulsed by the topic? I will argue that they believe that condoms limit their freedom and stymie their natural instinct. They are reluctant to talk about condoms because they cannot boast about them.

RESISTANCE AGAINST CONDOM USE

When I asked clients the reasons about their resistance to condom use, they commented that wearing condoms is like "taking a shower with a raincoat on" and "scratching an itch outside the boots." Complaints about the reduction in pleasure have been reported in Thailand, Cameroon, and Tanzania (Calves 1999; Knodel and Pramualratana 1994; Plummer et al. 2006).[9] Although discomfort and a lack of pleasure constitute part of the reasons for their abjuring condom use, I argue that a major reason is that clients abjure condom use as a political statement to rebel against the state repression of sexuality.

In *Red Lights: The Lives of Sex Workers in Postsocialist China* (T. Zheng 2009), I analyzed the clients' metaphoric use of *liang* (grain) in their reference to having sex with their wives as "*jiao gong liang*" (turning in the grain tax). I detailed how the history of the hierarchical system of food rationing in colonial and Maoist Dalian placed the Japanese and Maoist states at the top of the food chain and the men at the bottom. Meanwhile, men perceived their masculinity as lost in the alliance of the Maoist state and Chinese women to liberate women. Thus, semen and food were the pivotal points for symbolizing state power and social relationships. In the post-Mao era, men's attempts to recover their lost sexual identity with free-ranging promiscuity in the karaoke bars were further curtailed by the continued presence of socialist moralities and state laws. Thus, clients analogized turning in their *jing* (semen) to their wives as peasants turned in the grain tax to the state. The clients/peasants perceived themselves at the bottom of the hierarchy vis-à-vis the wives/state. I argued that the greater strain the state executed on the clients' sexual function, the more likely an economic view of the body emerged. The clients operated on an economy of scarcity where the semen was perceived as finite. The social stress of a lack of control was expressed in the bodily symbolism in which they assumed more control over their "limited" amount of semen. Self-perceived as the managers of their bodily assets, the clients exercised what I called "misappropriation"; that is, they allocated their semen between their wives and hostesses. Such a misappropriation of their *jing* (semen) was a mode of resistance, just as *liang* (grain) was misappropriated by peasants who rebelled by cheating the government of their taxes. The clients' subversive misappropriation was intended to maintain their bodies' independence as "impermeable, inviolable entities" (Brownell 1995, 243).

This theme of clients' resentment and defiance against state control of sexuality and pleasure continued throughout my current research. Clients expressed their resentment toward the cultural understanding of sex as evil and the state's need to restrict and repress it. This echoes the traditional cultural values of "sexual silence" between men and women in the Latino culture (Carrillo 2002; Gomez and Van Oss 1996).[10] As with Latino people who have demonstrated a strong favor for cultural change, clients in my research also vehemently voiced their dissatisfaction with this cultural norm. Client A said, "Sex is sin in China. We are taught that sex is the first and foremost evil among the ten thousands of evils. This Confucian idea is still prevalent in people's minds." Client A said that he has never talked about sex with his wife because he was too embarrassed. Other clients told me the same story. Client B said, "In China, no one likes to talk about it [sex]. Even with male friends, no one talks about sex. Sex is still a taboo. I don't want to answer these questions about condom use. I'll just say 'I don't know' to handle these questions, not because I don't know, but because I don't want to answer them." I thanked him for telling me honestly his resentment toward the questions on condom use. I then pressed him and said, "I understand that sex is a taboo and no one likes to talk about it, but why did the male clients respond to this topic with so much resentment? Why is the topic of condom use so unpleasant to them?"

Client B responded, "This is because of the conflict between the mainstream which is conservative, and the personal and private sphere, which is liberal. Just like the saying that goes, 'One Muslim can eat full, but two Muslims have to starve to death.' One Muslim person can eat pork when alone and private, but two Muslims will have to starve because of the taboo. This is the conflict between the conservative mainstream and the liberal personal sphere."

Client B appropriated the expression about one full Muslim vis-à-vis two starved Muslims to illuminate the conflict between the personal space that is free and liberal and the mainstream that is repressive and shackling. By "mainstream," he meant the government that repressed expressions of sexuality. Under this constraining and suppressive political environment, he was unwilling to talk about condom use because it belonged to his private sphere. This private sphere was liberal. It included patronizing and sleeping with hostesses and striptease in karaoke bars and sauna places. Because it is in direct contradiction to the mainstream, it is censored, controlled, and condemned by the mainstream. One might be justified in questioning client B's response since he was more than willing to talk about sex when it allowed him to brag about his sexual exploits. It was only after the topic of condom use and HIV/AIDS was brought up that he was concerned with the distinctions between the public and private sphere and claimed the right of privacy.

Client B expressed his indignant hatred toward the government's hostile attitudes toward extramarital sex among hostesses and male clients and premarital sex among college students. He said,

> Police often attack sauna bars and karaoke bars to arrest those who sleep with hostesses unless the manager is a friend of the police. Some of my colleagues were dismissed from their positions because they were caught by the police sleeping with hostesses. College students can be dismissed from schools for engaging in sexual activities. It's only recently that college students have been allowed to get married—the policy has only started to open up recently. All these behaviors run against the mainstream and they are the target of expulsion and punishment. That's why in personal space people can be very open and liberal, but they're not willing to talk about it in public. It's because of this conflict. This conflict forces people to appear conservative and traditional.

This tirade made it clear that client B saw extramarital sex as an active rebellion against government repression. Client B's words directly challenged the state's repression of extramarital and premarital sexual practices. He repeatedly emphasized the conflict between the state's control of sexual expression and the individuals' vehement rebellions against the repression in their private spheres. Although male clients were condemned, fined, and punished by the government for sleeping with hostesses, they continued to enjoy sexual pleasure in their personal spaces liberated from the mainstream. He stated that such a hedonistic sexual pleasure represented a liberating space from the suffocating repressive state because no one believed in the state ideology any more.

There was a thought in the past—communism. It was communism that guided people to do everything. Today this thought is no longer useful. It is of no use in the society; it is of no use in the job. It lost its function as guidance. It lost its meaning. Because China doesn't have any religions like other countries. China doesn't even have the simplest principle of being kind to others [*yuren weishan*]. People lost their directions and embraced all the new things. I lack an understanding of these new things. All I know is that the basic nature of human beings is evil [*renzhichu, xingbene*]. I have this sexual urge; I need sexual satisfaction; and hostesses can offer it.

This attitude, while not universal among the clients I interviewed, was representative of a large proportion of those who were willing to speak on the subject. Client B's words vehemently criticized the state ideology as useless and meaningless and pinpointed the detrimental effect of such an empty state ideology on people, that is, a moral vacuum and a lack of basic kindness to others. Such a moral vacuum drove clients to explore their sexual urges and pursue their sexual liberation as a voice of defiance and rebellion against the deceiving and repressive state ideology.

Client B's words encapsulate my argument that the state's repression and surveillance of individual private spheres instigated the clients' resistance through sexual liberation. Such sexual liberation not only included engaging in illicit sex with hostesses but also entailed resentment against condom use.

Hyde argues that, in competing with the state family planning apparatus, the market economy opens the door to new kinds of sexual practices (Hyde 2007). Certainly, the market economy has unleashed and catapulted male clients' sexual revolution. About condom use, Hyde argues that purchasing condoms in the market, compared with acquiring condoms from the state, is more discreet and increases privacy, which helps individuals resist state intrusion into their personal lives as "everyday practices of resistance" (Hyde 2007, 160, 168). In this particular aspect, male clients and hostesses in my research told me a different story. As I related earlier in the introductory chapter, some male clients who purchased condoms said that purchasing condoms in the market is a "public act," involving a great embarrassment. For instance, one client explained, "I am afraid of buying condoms. When I do, I drop my head like a criminal who has just committed a crime, hastening to escape after paying for the condom." Young unmarried women, such as hostesses in my research, encountered stigmatizing comments from the sales workers who called them sex workers. Although purchasing condoms in the market is embarrassing, they still preferred the market to the state apparatus because, as they told me, the state-distributed free, poor-quality condoms (too dry and easily broken), in limited quantities (only ten free condoms a month), and for married couples only. The difference between Hyde's research findings and mine could result from our different research locations and disparate research populations.

While Hyde concludes that the market "provides a potent weapon against STDs and AIDS through the sale of condoms" (Hyde 2007, 152), the story of male clients in my research points to a cultural dimension that thwarts and undercuts the potency of this "weapon." I define this cultural dimension as the perception of condoms by male clients. To clients, condoms symbolized state repression of sexual pleasure and state intrusion into the individual private sphere. By talking about condoms, restrictions are put on male clients' sexual pursuits, which help the state intrude into their private sphere. Talking about condoms limits their pursuit of pleasure. It reminds people that the state interferes with and represses their sexual desires. Hence, many clients regarded rejection of condoms as a political act of defiance. Complete sexual freedom for clients means symbolically reestablishing their control over their lives. I have argued that the post-Mao era is heavily defined by the Maoist era, the attempt to create complete equality between men and women brought a sense of emasculation among men (T. Zheng 2006). During the post-Mao era, men rebelled against emasculation by attempting to exercise complete sexual freedom. In the process of acting out this freedom with hostesses, they sought to achieve masculine liberation.

PURSUIT OF ABSOLUTE SEXUAL PLEASURE

Throughout my research, clients related that men strived for the feeling of absolute sexual pleasure without condoms. For instance, client Li said that ever since he was married, he used condoms only once. He said, "Many men are like me. We have never used condoms with women. If all of a sudden, you ask them to start using condoms, they will compare the feelings with condomless sex and feel really uncomfortable. They don't like the feeling and they will not use them any more." Li told me that he only tried condoms once with his wife to see what condoms felt like. He had to take it out during the middle of intercourse because it did not feel comfortable at all. I pressed him to explain; he said, "How can it be compared with flesh-to-flesh contact? Flesh-to-flesh contact gives stimulations. The stimulation is lost with a layer of barrier. Plus, the condom is oily and it just doesn't feel good to wear."

Client Li's answer is representative of my interviews with clients in general. In fact, during my interactions with the clients, there was a consistent embrace of hedonistic pursuit of pleasure. Clients' quest for absolute pleasure resonates with men in rural Tanzania (Plummer et al. 2006), who complained that condoms dull sexual sensation and defeat the purpose of sexual encounters. They aspired to achieve multiple ejaculations within one sexual rendezvous, both maximizing their pleasure and demonstrating their prowess (Plummer et al. 2006).

I will relate a story that I encountered during my research. Clients always told me how much they enjoyed the night shows at sauna bars. I was curious to know what kind of shows and environment they regularly enmeshed themselves in. I made friends with two clients from my research and asked whether they could take me to a night show at a sauna bar. They agreed, and took me to the most popular sauna bar in the city. At this bar, every night performances ran from 10 p.m. until 1 or 2 a.m. We came in around 9 p.m., went to the rest hall, laid on the sofa, and waited for the night show to start. Shortly after we sat down, two hostesses came along and asked A and B whether they needed Korean-style bone-loosening massages. A asked, "What is it for?" The hostess replied with a seductive smile, "You know, the kind of massage that helps solve your problem!" Clients A and B murmured to the two hostesses; I assumed that they were negotiating prices. After a while, A and B left with them to two separate private rooms. They were gone about twenty minutes; during that time, I looked around and noticed that I was the only female in the audience. By the time A and B came back, the show had already started. The show included a striptease dance and a male and female comedian. The content of the comedy show was blatantly sexual and antithetical to the mainstream morality. The show started off with the male performer telling the audience that he and his father shared his wife—the female performer. He commented, "My wife is very beautiful. My father and I share her every day, to save money. Other men also salivate for her, such as the emcee, although the emcee is eighteenth on the waiting list." The female performer said, "My body is very beautiful. My body is very white. I have a bushy pit inside of my snow-white and big buttock. It's deadly beautiful. If you have money, you can throw money inside my pit." She then started moaning to the emcee, and said, "My B [vagina] is so tight. Let me squeeze you dead. [repeat] Let's play." She asked the male performer, "Do you have money? I will play with you if you have money. If you don't have money, who wants to play with you!?" She then stretched forward the lower part of her body to the men in the front rows, calling them "bro" (*gege*) and moaning repeatedly, "I want it, I want it more." The male performer quipped, "Look at how horny you are. These men all want to undo their pants and put their things into your B [vagina]." Then he turned to the audience and said, "You can sleep with my wife."

The most striking aspect of the performance was the rebellion against mainstream morality. Mainstream morality indoctrinates people to satisfy sexual urges within marriages and to maintain premarital and extramarital chastity. Both marital purity and incest taboo were trumped, ridiculed, and superseded by an absolutely commercial mentality. The male performer told the audience that both he and his father had intercourse with his wife to save money. His wife told the audience that if her husband did not have money, she would not sleep with him. She was willing to sleep with any men who had money, regardless of how many and who they were. Their performance told the audience that the incest taboo and marital purity are meaningless and worthless compared with money. No doubt we should all forgo them for the sake of money and for the hedonistic pursuit of pleasure. The blatant sexual content of the performance was designed not only to provide sexual pleasure to the audience but also provided that sexual pleasure through enjoyment of the violation of conventional taboos.

As the show went on, the emcee urged the audience to pay 20 yuan to draw a lot that awarded a gift of 500 yuan. He repeated these words again and again to the audience, "Let's just play! Play is happy [*kaixin*], play is complete [*tongkuai*], play is absolute [*xiaosa*], play is stimulation [*ciji*]! Let's sell our children and buy a monkey to play with! Let's dismantle our houses and look for a cricket to play with. Let's just play!"

These words convey an outright rejection of mainstream morality and an absolute embrace of hedonistic pleasure, free from moral constraints. Mainstream morality tells people that family values should be primary. Observing family values entails taking care of children and maintaining a home. It is expected of any moral person in the society. Here the emcee inverts this mainstream morality and proposes a completely opposite moral view. That is, "Let's sell our children and buy a monkey to play with. Let's dismantle our houses and look for a cricket to play with." Play is the ultimate goal. Play is the final purpose. Not families. Not values. When you play, you have to play to the extreme, play to the absolute, play to the complete. But before you can do all that, you have to first relinquish all the mainstream constraints from your life and get them out of your mind. As he said, you have to be able to sell your children and dismantle your house. Only after you destroy these things that are emblematic of the shackles and the chains on your body and soul can you attain total freedom.

In this cultural environment, the clients are urged to reject what society tells them is important and relentlessly annihilate all that symbolizes mainstream values. That is, embrace an extreme and absolute pleasure and renounce everything imprinted or tinged with state-inculcated values. Condoms fall into such a category. It symbolizes state repression and control of sexuality. The dissemination of AIDS information is regarded by the clients as just another strategy that the state uses to scare them and control their sexual behaviors. To them, it is a fraud. Hence, they tend to reject the use of condoms. Wives also fall into such a category. As clients told me in the interviews, "We can find sexual pleasure with hostesses. Wives are the microcosm of the family. As everyone says, you get tired of her even if she is beautiful (*shenmei pilao*). No matter how good and how beautiful she is, she is always a barrier to pure pleasure. Eventually, you lose your feelings for her." Hence, they tend to engage in illicit sex with hostesses to pursue the utmost sexual pleasure.

The clients rejected the state's repression of sex and the entrenched cultural values and heralded a hedonistic view of free love and free sex. The intensity of the comedic performance and the testimony by the emcee suggests that while the clients are trying to free themselves from these entrenched cultural values, these values still have a hold on them that requires a resistance through protestations and satire.

Clients' rebellion against mainstream morality in urban entertainment sites resonates with previous researchers' findings of resistant strategies and challenges against ascribed sexual roles in illicit spaces such as streets, karaoke bars, and massage parlors (Hubbard 2000; Sanders 2004). Shen, for instance, depicts the formation of transgressive sexual liaisons between traveling Taiwanese businessmen with Chinese women through bars and nightclubs in mainland China (H.-H. Shen 2008). Ho Sewe Lin, in his study of Japanese love hotels, demonstrates how transitory moments of individual freedom are realized through commercial love hotels that offer rooms primarily for private sexual activities (H. S. Lin 2008). As Ho argues, free sex in love hotels represents a site of resistance and contestation against socially ascribed rules that confine sex within conjugal relationships, for the purpose of reproduction. Individuals' liberal sexual behaviors in the public sphere represent their agency in pursuing individual fulfillment, emphasizing pleasure and liberation, and defying the collective-based society. In a similar vein, Williams and Lyons (2008) also illustrate how a commercial sex information Web site operates as a free discursive space for Singaporean men to challenge the state's demand of a heteronormative expression of masculinity. These alternative "illicit" sites act as "borderland spaces between the mundane and the extraordinary" and challenge the "ideas of morality [that] have served to naturalize the view that sex must be based on an exchange which is meaningful both materially and emotionally and is based on procreative sexual intercourse" (Little 2003, 403; Pritchard and Morgan 2006, 764; but also see Huang and Yeoh 2008, 4–5).

In my research, clients rejected AIDS and believed that inculcating fear of AIDS is a strategy the state uses to curb their free sexual expressions, which threaten the state. This sentiment is analogous to the gay community in their sexual revolution during the early 1980s. When little was known about AIDS, the gay community in the San Francisco area was unreceptive to attempts by local government and medical institutions to promote safe sex or condoms and to discourage high-risk behavior like anonymous sex in bathhouses. These educated efforts were largely thwarted by the history of misunderstanding and outright prejudice that had existed between the gay community and the American medical establishment. Before the AIDS outbreak and outreach programs, homosexuality was defined by the medical establishment in general and psychiatry in particular as mental illness. This view of sexual orientation as a pathological behavior caused homosexual groups to distrust government and medical agencies so that even Foucault, a famous French intellectual, believed that AIDS was designed to further stigmatize homosexuality. It was not that Foucault did not have the facts, but rather, it was the inseparable relationship between medicine and, in this case, politics, history, and power that led to the unchecked spread of AIDS in the early period and the tragic loss of life.

Clients held similar kinds of resentment against the state's regulation of their pursuit of sexual pleasure and expressed an outright hostility to HIV/AIDS and condoms. To them, these are the state's tools and weapons to control and police their sexuality. AIDS and condoms represent exactly what they rebel against—control and regulation of their sexual pleasure. Warning about AIDS and condom use shackle their sexual pursuits and thwart their absolute enjoyment of free sex. They reject mainstream morality and constraint and embrace sexual freedom.

"VALIANT" CLIENTS FEARLESS OF STDS

One of my interview questions was, "What do you think of your male friend who does not use condoms with hostesses?" The common response I received was, "I think he is very valiant (*meng*). You know, very brave, fearless, has the courage to take risks." Their responses showed that clients attempted to project a brave and valiant (*meng*) persona for the peer group through abjuring condom use.

Here, the idea of "valiant," or "virile," is similar to the meaning of masculinity in other cultures, such as in Africa and in Latin America. Researchers have pointed out that, in South Africa, condom use is seen as undermining South African men's notions of masculinity that valorize flesh-to-flesh sex with multiple sexual partners (Holland et al. 1994a, 1994b; McGrath 1993),[11] fertility (Abdool et al. 1992), and pleasure (Preston-Whyte et al. 1991). As researchers indicate, masculinity in Southern Africa is tied to flesh-to-flesh sex (Wojcicki and Malala. 2001; Webb 1997). As Campbell argues, workers view it as necessary for a man's good health to maintain balanced levels of blood/sperm within the body (Campbell et al. 1998). Informants spoke of the way in which the buildup of sperm led to a range of mental and physical problems. Flesh-to-flesh sex was regarded as the only pleasurable way of meeting male sexual desires, with condoms seen as cold and unpleasant. Masculinity, associated with physical strength and bravery, served as a key coping mechanism whereby miners deal with the harsh and dangerous working conditions of underground mining (Campbell et al. 1998, 52). South African men who refuse this dominant masculine notion by using condoms or avoiding sex are taunted, teased, and ridiculed as stupid (MacPhail and Campbell 2001). Rural Tanzanian men reject condoms and believe that conception defines masculinity (Plummer et al. 2006). Researchers have also observed the pervasive *machismo* culture among Latino men that defined machismo behaviors as having multiple sexual partners, imposing physical abuse and restrictions on women's mobility, and so on (Inciardi et al. 2000; Singer 1990). Being *macho* also means being passionate and irrational during sexual intercourse and not disrupting the sexual flow with condom use (Carrillo 2002; Ibañez et al. 2005; Martin and Gomez 1996).

Among the clients in my research, a similar concept of machismo is manifested through engaging in illicit sex with multiple sexual partners, not wearing condoms, and not fearing STDs, while imposing control over hostesses. Previous studies in South Africa and Thailand have shown that peer norms are powerful in affecting group behaviors (Campbell et al. 2001, 1614; Maticka-Tyndale et al. 1997). Peer norms can either facilitate discussions of safe sex and generate safe sexual behaviors or function to promote unsafe sexual behaviors and encourage risk (Campbell et al. 2001). For instance, research on Thai men showed that peer pressure prompted them to participate in risky activities during group partying and consumption of alcohol (Maticka-Tyndale et al. 1997). In my research, peer norms among clients valorize macho display of a fearless spirit and the pursuit of absolute sexual pleasure and prevent them from protecting their health.

Unafraid of STDS

Research on sexual partners and risk patterns for HIV infection in China has revealed that if male sex consumers held the belief that few or no STDs could be cured, they would be more inclined to engage in consistent condom use (Parish and Pan 2006). Knowing the correct information that most STDs can be cured would only lead to less condom use (Parish and Pan 2006, 199). Do clients in my research believe that few or no STDs could be cured? How do clients perceive the risk of STDs? When pursuing absolute sexual pleasure, do clients fear STDs and HIV/AIDS?

When I asked client A if he was afraid of getting STDs, he responded, "Not at all." "Why?" I asked. He smiled and said, "Well, even if you just drink cold water, you may get something trapped in your teeth. Not every hostess has disease. If it just so happens that the hostess you sleep with has disease, you have to acknowledge that you have had bad luck. You'll have to say, well, I happened to meet such a hostess. I have nothing to say but to recognize my bad luck!" Similarly, client B said, "Not every hostess has disease. Every day you see other people sleeping with hostesses without condoms and they have not caught any diseases. You know that it's your bad luck if you catch it once in while."

To the clients, catching diseases from illicit sex is a matter of luck and chance. A man should be brave enough to take risks and show a fearless attitude toward STDs. Clients asserted that they did not care about STDs because so many people got them and one shot was enough to get treated. Client C said, "Did you see how many people queue up in the municipal STD hospital every day to be treated for venereal diseases? Tons of people! So no one feels embarrassed about having venereal diseases." "All my friends had venereal diseases because they just did not care," he told me, "but there is nothing to worry about because they are all treatable.

My friends go to the hospital regularly and receive shots once a month. One shot makes them well immediately. That's all it takes—one dose of medicine or one shot, the disease is gone. So it's very easy to be treated."

This dialogue demonstrates the need for these men to conform to this new rebellious group's identity, which defines them as real men. Whatever these men's real fears might be, they are attempting to project a certain kind of masculine image to me as an interviewer. Clients in my research tend to take venereal diseases lightly because it is not only easily treatable but also something shared by their male friends in their environment. As they expressed throughout the interviews, a valiant man is not timid; a valiant man does not worry about taking risks; a valiant man is fearless of STDs. In fact, as client C told me, his friends "took it so lightly that they were even trying to hook up with hostesses in the STD hospital when they were queuing to see the doctor."

If clients believe that taking risks is a mark of their virility and valor, then what about HIV/AIDS? When I asked the clients this question, they immediately brushed it away and responded, "AIDS in China is mostly transmitted through blood transfusion. Not through sexual intercourse. It's other people's problem, not mine."

POWER OVER WOMEN

The valiant man is not only defined as unafraid of STDs but is also marked by his control over women. This control includes not using condoms when engaging in sex with hostesses and wives, inflicting pain on the bodies of hostesses, and deeming it women's responsibility to use contraception.

INFLICTING PAIN ON HOSTESSES

During my discussions with owners of adult health shops, clients, and hostesses, I heard many stories of clients deliberately inflicting pain on hostesses to exercise their power. Hostess Mei relates this story. Mei met a client at her karaoke bar and slept with him that night. A couple of days later she ran into him on the street, and he asked whether she could accompany him to a banquet with his male friends. She was in a hurry to see a friend, so she patiently explained to him that her friend was sick and that she had promised to attend to her. She apologized to the client profusely and promised that next time she would definitely accompany him. Several days later, the client called her and asked her to meet him at a hotel. She immediately went. In the hotel, the client penetrated her without a condom. She said that the client must have taken some yang-strengthening medicine before hand because he did not stop until several hours later. She was exhausted. He finally stopped, took out a brand-new 100 yuan from his wallet, folded it into the smallest piece, and inserted it into her vagina. The brand-new 100 *yuan* is a much bigger, sharper, and scratchier piece of paper than the U.S. dollar. After it was inserted, it expanded and exploded in her vagina, causing such a huge hemorrhage that the blood soaked the entire bed sheet. She yelled in excruciating pain, only to hear the client saying coldly, "Just so you know, next time, never mess with me!" He left the room. She could not move because of the sharp pain. Later she called the emergency and was sent to the hospital. She had to go through a series of surgeries in her vagina and was lying in bed for several months before she recovered.

After this event, the client proudly disseminated the story to his male friends, and everyone shared a good laugh about it. His male friends said, "That's great! You taught her a big lesson! See if she dares to mess with you again!" Among his peer group, he established himself as a brave and valiant man with a misogynist and violent attitude toward women, manifested by his penetration without condoms and his willingness to inflict pain on the hostess.

WOMEN'S RESPONSIBILITY

As illustrated, the other element of being a valiant man is to hold women responsible for contraception. During my research in the karaoke bars, every hostess had had abortions from one to ten times. Clients' wives also commented that their husbands did not use condoms because they only cared about their own pleasure and not women's suffering. Almost every wife I interviewed had had abortions twice or more because their husbands did not like to use condoms. Some experienced deterioration in their health situation after the insertion of an IUD. Once they took it out, they got pregnant again and had to have more abortions.

In China, the belief is that birth control pills harm women's health (see also Goodkind 1997), so condom use is considered the benchmark of a caring husband. "Valiant" men, however, pursue absolute pleasure and exercise their power through penetration without condoms. In an interview, a male client responded to the question of condom use, "There is no feeling. Besides, Chinese girls are very clean and I can always make them get an abortion" (Matuszak 2003). In early 2003, China's State

Family Planning Commission publicized that men purchase an average of three condoms per year (Matuszak 2003). When pregnant, abortion is "a quick and relatively cheap solution" (Matuszak 2003). It is estimated that between the ages twenty and twenty-nine, 27.3 percent of the women have had abortions (Wei Wang 2005). An abortion clinic in Sichuan averages fifteen abortions each day between 8 a.m. and 12 p.m. (Matuszak 2003). The Dalian local newspaper encouraged women to use IUDs for contraception because the pain of an IUD is much less than that of abortion (Wang 2005). Yan Che and John Cleland, in their research on contraceptive use after marriage in Shanghai, reported that the IUD was the preferred method for most couples by the third year after childbirth (Che and Cleland 2003).

In my questionnaires, only 2 percent of clients responded to the question "Where do you purchase condoms?" In my subsequent interviews with the clients, I asked who should be responsible for providing condoms when they slept with hostesses. Clients responded, "Of course it's the hostesses' responsibility. They use their mouths to put on condoms for men." Client D said, "Men don't buy condoms, women buy condoms. Men buy yang-strengthening medicine [*zhuangyan yao*] and aphrodisiacs [*cuichun yao*]."

Yang-strengthening medicines were supposed to help lengthen the intercourse period, and aphrodisiacs were used to arouse hostesses to maximize men's sexual pleasure. I went to the adult health shops located near the karaoke and sauna bars and checked out these yang-strengthening medicines. They were called Wolf No. 1 (*lang yi hao*), From the United States: Erect for Three Days, Trumping Viagra: Virility and Dominance (*xiong ba*), and Golden Viagra (*Jin Weige*). The instructions claimed that these medicines employed high technology from the United States to extract the essence of traditional Chinese herbs. The advertisements on the packages claimed to keep a man erect for up to seventy-two hours. The shop owner told me that clients came in to buy these medicines before they went to karaoke bars. Since some medicines produced instant effects, whereas others were effective an hour later, clients purchased the appropriate medicines according to when they planned on sleeping with hostesses. They also purchased aphrodisiacs for hostesses. The shop owner handed me the most popular aphrodisiac among clients that was labeled Temptation Candy. The package was covered with two pictures: a picture on the front of a naked white woman with seductive postures, and a picture on the back of a naked white woman with a thick snake around her body. The snake here is emblematic of the temptation and seduction of women. The package said, "One candy can take you to the pinnacle of happiness."

As the clients and shop owners told me, men purchased aphrodisiacs to seduce women, and the yang-strengthening medicines were to achieve the maximum sexual pleasure for men. It is the hostesses' responsibility to purchase condoms and put them on for men. This act should not be interpreted as part of the titillation of foreplay but, rather, as symbolic of the subservient role of women and the contempt of men for condoms.

This was also confirmed by my interviews with the general manager of the Dalian division of the UK Jissbon Global Company. The general manager said that, over the years, mostly women bought their products. So they changed their tactics and started marketing to women. She said, "Jissbon Global Company has been established in Dalian for two to three years, and our products are sold in supermarkets, drug stores, and convenience stores. In the past, our product was specifically designed for men. As you see, our emblem used to be an animated condom with a face of a man. Now we have changed our strategy and designed our products for women instead. We target women because it is mainly women who buy our product. Now the emblem is more conservative and more subtle. It has flowers and grass."

Although condom companies have switched their packaging strategies to cater to women's taste, it is still culturally denigrating and disparaging for unmarried women to purchase condoms. Unmarried women buying any kind of contraceptive are usually subject to shame and rumor. Hostesses tend to buy condoms in the adult health shops close to karaoke and sauna bars, and shop owners offer them special kinds that are more durable for drunk or violent clients. When women other than hostesses purchase condoms, however, they are ridiculed and despised as hostesses. In an interview conducted by the *New Weekly* journal about condom advertisements, a twenty-three-year-old married woman told a story of humiliation when she tried to purchase condoms at a state-owned drug store. At the store, she asked the female shop assistant to recommend the best brand. The shop assistant looked her up and down and said, "You don't know which kind is good?" The customer said, "No, that's why I am asking you." The shop assistant scoffed and said, "You should ask your colleagues or your male clients!" It then dawned on her that she was taken as a hostess.

Until recently, China's mass media have portrayed women carrying condoms as sex workers (Qu et al. 2002; Xin 1999, 1422). Research in Cameroon and South Africa has also shown a stigma attached to women carrying or purchasing condoms (Calves 1999; MacPhail and Campbell 2001).[12] The social and cultural beliefs are that women who carry condoms actively seek or plan sex. Hence, they are associated with negative reputations and labels. Such social milieu creates a predicament for women to procure or purchase condoms, as it pressures women to only engage in sex within the confines of serious and trusting relationships and hence nonuse of condoms (Hillier et al. 1998; Holland et al. 1990, 1991; Ingham et al. 1991).

On the one hand, the "valiant" clients deem it women's responsibility to purchase condoms. On the other hand, women purchasing condoms are subject to humiliation and are despised as hostesses. This situation makes it difficult for women engaged in illicit sex who are not hostesses to protect themselves. As a result, as I mentioned, hostesses felt more comfortable purchasing condoms at nearby adult health shops because shop owners would be less judgmental and less harsh because these shops depend on hostesses and clients as their mainstay consumers. Also, positioned close to karaoke bars and sauna bars, they are more familiar with and accustomed to the sex culture in the surroundings and, hence, more accepting of it.

CONCLUSION

There is a discrepancy between knowledge about the efficacy of condoms and the refusal by many clients to use them in the city of Dalian. While 77 percent of the general Chinese population did not understand the protective value of condoms, among the client group I interviewed, comprised predominantly of well-educated government officials and entrepreneurs, 100 percent understood that condoms offered protection against sexually transmitted diseases, including HIV.

I sought to resolve the conundrum of the incongruity between the clients' knowledge about the efficacy of condoms and their refusal to use condoms. I argue that the reason for unprotected sex lies in clients' perceptions of condoms in their social and cultural context.

Three major cultural and political factors contribute to the gap between clients' medical knowledge and their refusal to use condoms. These factors include (a) the rejection of state control over natural sexual expression, (b) the perception of nonuse of condoms as representing a kind of bravado (*meng*) as perceived by the peer group, and (c) the perception of contraception as the woman's responsibility.

What my research shows is that the condom has a special symbolic resonance with clients. Because condoms symbolize state repression of sexual pleasure and state intrusion into individual private spheres, clients sought to rebel against the restrictions on their sexual pursuits and achieve masculine liberation. Such sexual liberation was multilayered. It not only entailed resentment against condom use and engagement in illicit sex but also necessitated complete flouting of mainstream moralities and pursuit of absolute sexual pleasure. It is also characterized by a carefree posture in the "valiant" male culture that regards clients' own pleasure, virility, and control over women rather than the welfare of women. While we have noted similar male cultures in Latino and African society, what seems to be unique in China is its rejection of a historically controlling government, which was particularly severe during the Maoist era. While clients are experiencing a new freedom through their symbolic rebellion against state control, the cost of their outright sexual liberation is the health of women.

Chapter 5

Perceptions toward Condom Use among Dalian Hostesses

Introduction

Since 1978, the state's proconsumption stance has opened the way for the reemergence of nightclubs and other leisure sites. To avoid any residual negative connotations from the Maoist era in which nightclubs, dance halls, and bars were classified as emblems of a nonproletarian and decadent bourgeois lifestyle, nightclubs in the current post-Mao period were referred to as karaoke bars, karaoke plazas, or *liange ting* (literally, "singing practice halls"). These new consumption sites were prominent in the more economically prosperous special economic zones (SEZs; Jian 2001). Patrons were mainly middle-aged businessmen, government officials, policemen, and foreign investors. Clients could partake of the services offered by hostesses and, at the same time, engage in "social interactions" (*yingchou*) that helped cement "relationships" (*guanxi*) with their business partners or their patrons in the government (G. Wang 1999). Hostesses played an indispensable role in the rituals of these male-centered worlds of business and politics.[1]

The hostesses or escorts who worked at karaoke bars were referred to by the Chinese government as *sanpei xiaojie*, literally, "young women who accompanied men in three ways." These "ways" were generally understood to include varying combinations of alcohol consumption, dancing and singing, and sexual services. These women, mainly seventeen to twenty three years old, formed a steadily growing contingent of illegal sex workers. Their services typically included drinking, singing, dancing, playing games, flirting, and caressing. Beyond the standard service package, some hostesses offered sexual services as requested by clients for an additional fee.

Most of these hostesses were rural migrant women. Dalian's rapid growth had made it a magnet for labor migrants. By 1998, the most conservative estimate placed the floating population in Dalian at around three hundred thousand. Institutional (i.e., household registration policy) and social discrimination have forced the most of these migrants onto the lowest rung of the labor market.[2] Migrants commonly work as construction workers, garbage collectors, restaurant waitresses, domestic maids, factory workers, and bar hostesses. Many female migrants find employment in Dalian's booming sex industry. The city police chief estimated in 2001, that 80 percent of the total population of migrant women worked as hostesses in the nightclub industry. He might have exaggerated, but this figure suggests that a high percentage of migrant women work as bar hostesses.

In the entertainment industry, the constant interactions between hostesses and clients often led to a nuanced and ambiguous relationship, wherein power, control, romance, and manipulations were intricately intersected (T. Zheng 2009). The complexity of their relationship is crucial to understanding negotiations of condom use and risky sexual behaviors.

As noted in the introductory chapter, the political and economic factors driving the HIV epidemic are intertwined with gender hierarchy and sexuality, wherein women, and low-income women in particular, are vulnerable to HIV infection. The previous chapter on male consumers' resentment of condom use and pursuit of absolute pleasure without barriers provides an essential background to understand the cultural pressure women face in sexual decision making. How do hostesses deal with this cultural pressure? What are their attitudes toward sexually transmitted diseases (STDs) and HIV/AIDS? Do they have the power to determine condom use or nonuse? Can hostesses convince male consumers to use condoms? How do they cope with clients who tend to abjure condom use and resort to unprotected sex? Are they complicit in the unprotected sex, or are they resistant to it? What kind of political economic factors are at work to shape patterns of sexual practice? Do these factors facilitate or impede hostesses' self-protection?

This chapter provides an ethnographic understanding of the political economy of HIV/AIDS in the urban Chinese sex industry. As explicated in the introductory chapter, I define political economic forces as political policies, economic differences, gender hierarchies, and power relations.

This chapter investigates elements of the sex industry that contribute to condomless sex such as competition between women for clients and violence in the industry that structures sexual practices and contribute to noncondom use between male sex consumers and female sex workers.

ATTITUDES TOWARD STDS AND HIV/AIDS

During the summer of 2005, I shared an apartment with two hostess friends.[3] One got pregnant with her client lover and had just had an abortion. She spent days just lying in bed to recuperate from the abortion. One day her hostess friends came over to see her and brought eggs and fruits, which were culturally defined as nutritious food for recovery from abortion.

While we were hanging out in the apartment eating, I took out several AIDS booklets obtained from the local Centers for Disease Control and Prevention office, and gave everyone a copy. I suggested to them, "Let's read this AIDS booklet together." I was a bit surprised at their disinterested responses, "I don't feel like reading anything. I am too exhausted to read." I took over the booklet and started reading the part on sexual means of transmission of AIDS. My reading invited protests from the hostesses, "OK, stop! I don't want to know this stuff. How do you expect us to continue living after knowing all this? We are better off not knowing about this because otherwise we can't continue living." One hostess asked, "Who would pay for this disease? How much would it be?" I read, "The country will pay for the drugs free of charge." Hostesses immediately replied, "If you get it through donating blood, of course, the country will provide drugs for you for free. If you get it through sleeping around [*guihun*], there is no way the country will provide free drugs for you!" That ended the conversation on AIDS.

Hostesses' responses revealed their negative attitude toward HIV information and their mistrust of the government. Why did they reject HIV information? Why did they show such strong distrust of the government? Understanding these two questions would require understanding the cultural context of hostesses' lives. Hostessing is considered an illegal profession, and hostesses have always been the subject of police arrests, heavy fines, and random imprisonments (Pan 2007; T. Zheng 2009). Chinese sociologist Pan Suiming has pointed out that the government's ban on hostessing is unreasonable because of the following three reasons (Pan 2007). First, the government has failed to offer employment opportunities for rural migrant women in the city. Second, the government has not only failed to award hostesses for the relief of the unemployment pressure but also arrested, fined, and harassed them. Third, the government has never explained why sex work cannot be regarded as a job (Pan 2007). A plethora of government-mandated crackdowns and random arrests made hostesses face legal antagonism, social discrimination, and intense exploitation and violence by owners, managers, and madams (T. Zheng 2009).[4] Living their lives in fear of police arrests and government crackdowns, hostesses are cognizant of their prescribed illegal, low, and marginalized status.

In addition to their marginalized status in society, hostesses' identities as rural migrants anchored them at a much lower status than urbanites in general and urban clients in particular (T. Zheng 2003; T. Zheng 2004). China scholars have written extensively on the second-tier citizenship of rural migrants and on the rural-urban apartheid system existent in current China (Solinger 1999; L. Zhang 2001).

Their dual marginal status as rural migrants and illegal hostesses has made hostesses well aware of their limitations on their social mobility. In other words, they do not have much control or power over their lives. There is only a micro-level niche wherein they can maneuver. The awareness of their lack of control in their lives has made hostesses ruthlessly cynical. They want to be oblivious of arenas where they have no control. They do not want to be reminded of their lack of control because it can only make their lives more miserable.

Knowledge about AIDS and sexual means of HIV/AIDS transmission fits in this arena. Dismissing the knowledge and wanting to stay ignorant is a consequence of their lack of control in their sexual relationships with clients. Indeed, as I show later in the chapter, it is mainly clients' decisions whether to use condoms.

Just as hostesses wish to stay ignorant about HIV, Hillbrow sex workers in Britain wish to be uninformed of their HIV status (Henrickson 1990). Researchers have observed that some Hillbrow sex workers have chosen not to test for HIV, and others chose not to return for the results because they were not interested in learning their HIV status. Both hostesses and Hillbrow sex workers chose not to know, to stay uneducated and uninformed of HIV knowledge or HIV status. In Hillbrow, sex workers fear that if they tested HIV-positive, their lives would be more miserable. They would be depressed and forced to leave work. Staying ignorant is their coping strategy to handle the stressors in their lives: poverty, violence, and discrimination in a tense, dangerous, and difficult world. In the case of hostesses, since they do not have control over condom use in their sexual relationships with male clients, information and knowledge about sexual means of HIV transmission would only contribute to their already high levels of stress.[5]

Indeed, because of various factors, including gender hierarchy and culturally prescribed sex roles, HIV sentinel surveillance with female sex workers found that, nationally, only 17 percent of male customers used condoms consistently (Qu 2001; Qu et al. 2002). This rate was extremely low compared with usage among female sex workers in other Asian countries: 48 percent in Vietnam (Thuy 2000) and 89 percent in Thailand (Mills et al. 1997). Consistent condom use in China was low primarily because clients refused to use them. Researchers across the globe have maintained that, because of the power hierarchy between male clients and female sex workers, clients are the key determinants in whether condoms are used during sex exchanges (Aral and St. Lawrence 2002; Elifson et al. 1999; McMahon et al. 2006; Vanwesenbeeck et al. 1993).

Indeed, hostesses are disadvantaged vis-à-vis male clients, not only because of power hierarchy but also because of economic dependency and survival strategies in a highly competitive environment.

COMPETITION FOR CLIENTS

In the coterie of hostesses, having a rich client lover marks a hostess's status. A client lover brings a hostess "good times" (*hao guo*), represented by wealth, a comfortable livelihood, and wide social networks. The crucial role of clients makes it a daily reality for hostesses to compete for clients. Hostesses are in constant competition with one another over their clients' status and the benefits they can glean from clients. Fierce competitions can even disintegrate hostesses' informal alliances and make them betray one another (T. Zheng 2008b).[6]

Means of competition vary from putting one another down in front of clients to stealing other workers' client lovers. Dissolution of hostess allies and networks as a result of stealing and snatching is common among hostesses. As a consequence, hostesses grow so distrustful of one another that no matter how close they are, they do not forget to guard against one another. Indeed, it is considered taboo to introduce your client lover to hostess friends. During my research in the karaoke bars, I was reminded again and again of this taboo: "If you do so [introduce your client to your hostess friend], the next day you will see him with your hostess friend."

At the karaoke bar, I witnessed many cases of hostesses seizing clients from one another, subsequently leading to intense fights among themselves. For instance, a regular rich client visited the karaoke bar and ordered the same hostess, Lin, for the night. During the time when Lin went to the restroom, Hong slipped into the karaoke room and sat next to the client, saying, "You are such a decent and classy guy. Why do you like a country bumpkin like Lin? She is from a rural village. Being with her is not appropriate for your identity. I am Dalian local, an urban girl. You can have me instead." She displayed her urban identification card to the client and put her arms around his. When Lin came back and learned the situation, she started a huge fight with Hong. The fight involved all the hostesses, resulting in a sharp division between urban and rural hostesses.

Even among rural hostesses, stealing is a common occurrence. Ling and May came to the karaoke bar from the same area as good friends. One day Ling was told that her client lover was with May at a sauna bar. Ling immediately headed to the sauna bar and caught the two of them together in the rest hall. Ling grabbed her client, laughing and talking to him, deliberately shunning and ostracizing May. From then on, the network between Ling and May was broken.

While previous researchers have emphasized networks based on common locale among migrant workers in urban China (Hershatter 1986; Honig 1986; Lee 1998; Ngai 2005; Rowe 1984; Strand 1989; L Zhang 2001), my research of hostesses showed that this network was trumped by their intense competition for clients because clients were the source of their survival, livelihood, and benefits (T. Zheng 2008b).

Research on female sex workers in Johannesburg has demonstrated similar competition tactics (Wojcicki and Malala 2001). Sex workers in Johannesburg deal with the competition by beating up other women, and either passing or explicitly saying negative remarks about the looks or behaviors of other sex workers to clients. In both cases, hostesses and sex workers in Johannesburg strive to satisfy clients' desires. They are fully aware that they must please clients, and they must be aggressive in their "game" to win with clients—to get them to give up their money. They deploy alcohol to loosen up and become more forward with clients. Environments in both places are competitive wherein women try desperately to seduce clients (Wojcicki and Malala 2001).

Like the sex workers in Johannesburg (Wojcicki and Malala 2001), the competitive environment wherein clients are critical for hostesses' survival has profound ramifications for condom use. In both places, in the competitive environment of sex exchanges for money, women find it difficult not to engage in condomless sex and capitalize on clients' reluctance to use condoms in sexual exchange (Wojcicki and Malala 2001). Like the situation of female sex workers in Johannesburg, attracting clients is a competitive business for hostesses. They are aware that some women may be willing to have sex at lower prices or have sex without a condom. They are also sensitive to competition from younger girls and other entertainment places. They perceive the sexual relationships with clients from a work perspective for economic gain. Like any workers in any job, they strive to be paid the most money possible for their efforts. If some women are willing to accept clients without condoms, then it is increasingly difficult for other women who require condoms to attract clients. As noted in Chapter 4, since many clients resent condom use and pursue absolute pleasure, not using condoms is common. It is not surprising to learn that some hostesses, like female sex workers in Johannesburg, choose to have condomless sex to cope with a difficult environment and to survive in a competitive market.

NONCONDOM USE AS A SURVIVAL STRATEGY

According to the hostesses, even though some clients would refuse condom use, they would still propose condom use with strangers. However, after first-time sex, clients were not strangers any more. Rather, they became embroiled in a lover relationship and relinquished condom use in subsequent sexual intercourses. Hostesses rationalized this change from request of condom use to rejection of condom use as imperative to enter a romantic and intimate relationship with clients to guarantee themselves a constant financial source and social network.

As I discussed in other articles, at the first meeting in the karaoke bar, hostesses used the term "husband" to address the clients who had chosen them (*laogong*; T. Zheng 2008a). They threw themselves into the embrace of their husbands. They sang romantic songs to their husbands. They played with their husbands. They danced with their husbands. In short, they performed all kinds of activities to achieve immediate intimacy with clients. Their goal was to have clients keep them as mistresses or marry them as wives for long-term financial security.

Each hostess had several client lovers. Some had client lovers and a boyfriend, such as hostess Li, for example. Her boyfriend was well aware of her client lovers but still encouraged her to glean economic profits from them. One day when we were hanging out, her boyfriend egged her on to get a new cutting-edge mobile phone from her client lovers. Puzzled by his positive reaction to her client lovers, I asked him, "Aren't you jealous of her client lovers?" He laughed and said, "No, I don't mind it at all.

When you are pressured for survival, you'll know that material things are much more important. I always urge her to cheat clients for more money, more cell phones, and more clothes."

Because of their dependence on clients for money, material goods, and social benefits, hostesses need to perform well to coax clients into a romantic and intimate relationship. In this quasi-conjugal relationship, clients treated hostesses as their wives and offered them financial support. One of the skills in luring clients into such a relationship was to engage in sex without condoms to establish trust and intimacy with clients and separate their private sex from commercial sex. In other words, a romantic relationship is necessitated by rejection of condom use, just like in a conjugal relationship. As indicated in Chapters 1 and 4, condom use was rare among married couples.

Research on female sex workers elsewhere has also demonstrated that sex workers exhibit low condom use with intimate partners to demarcate commercial from private sex and with clients to elicit more money to support their boyfriends (Macaluso et al. 2000; Van den Hoek et al. 1988; Waddell 1996b; Warr and Pyett 1999). Women are more likely to engage in riskier sexual practices with regular sex clients because of the more intimate nature of their relationships (Roche et al. 2005). For instance, female sex workers in Australia engage in condomless sex with boyfriends and husbands to demarcate work sex from non-work sex (Pyett and Warr 1997; Waddell 1996b). Condoms thus become a marker that separates commercial from intimate sex.

Indeed, it has been proven on a global scale that the nature of the relationship influences condom use. Condoms are much less acceptable by couples involved in cohabiting or romantic relationships, and noncondom use is justified by trusting relationships (Blecher et al. 1995; Carole Campbell 1995; Campbell et al. 2001; Cohen and Trussel 1996; Worth 1989). In fact, condom use decreases as a relationship grows more stable and more intimate over time. For instance, studies in South Africa, Cameroon, and Zambia have demonstrated that interviewees exhibit strong negative attitudes toward condom use in these kinds of relationships (Agha 1997; Calves 1999; Calves, Eloundou-Enyegue, and Meekers 2004; Maharaj and Cleland 2004). In Zambia, trusting partners is the most commonly cited reason for nonuse of condoms (64 percent; Agha 1997). In London, Manchester, Tanzania, and Khutsong, interviewees classified their new relationships as serious and incorporated issues of trust to justify their noncondom sex (Campbell et al. 2001; Holland et al. 1990, 1991, 1994a, 1994b; Klein et al. 1999).

Studies of sex workers in southern Africa (Wojcicki and Malala 2001) have also demonstrated that because regular clients occupy the fuzzy middle ground, sex workers do not use condoms with them. They state that although they may not take a stranger who desires sex without a condom, they will agree to condomless sex with regular clients, as regular clients provide a steady stream of income. Many sex workers feel that the more familiar they are with a client, the less risky it is to have condomless sex with him.

Like sex workers in southern Africa, hostesses also consider regular clients as a stable source of income. Hostesses' livelihood depends on their regulars. Indeed, many are kept as long-term mistresses. To demand safer sex with regulars could jeopardize the relationship. Therefore, despite their knowledge of sexual routes of HIV transmission, many hostesses still accept condomless sex with regular clients.

Hostess Lin had several male client lovers. One night she hooked up with another client. On their first night together, he gave her 5,000 yuan. He also used condoms that night. At their second time together, he quit using condoms and offered her the same amount after sex, telling her, "Don't go work in that place in the future. I'm keeping you. Here's the money for today." During the time when she was kept as a mistress, she was given at least 3,000 yuan a month, excluding all the leisure activities such as sightseeing at the seaside and scenic spots in the city. She stayed as his mistress at the apartment he rented for almost a year.

On their first night, the client used condoms with hostess Lin, symbolizing the commercial sexual relationship. Since the second meeting, however, with his rejection of condom use, the nature of their relationship was changed from commercial to intimate and romantic. In other words, nonuse of condoms was co-opted by hostess Lin as one of her tactics to secure her future financial source by turning him from a client into her client lover. At the same time, unbeknown to him, Lin was also kept as a mistress by several other client lovers. Over the years, hostesses have honed their skills to maneuver through different client lovers and to keep it secret from their lovers. For hostess Lin, with each lover offering her 3,000 yuan a month, she told me that her monthly income exceeded 10,000 yuan.

Other studies among female sex workers have similarly observed a significant decrease in condom use as relationship status changes from new to regular (Macaluso et al. 2000). In other words, the type of partners affects condom use. Condom use has been reported the lowest with regular clients and love mates (Basuki et al. 2002; Bloor 1995; Day 1988;

Morris et al. 1995; Oladosu et al. 2001; Tran et al. 2006; Walden et al. 1999; Wilson et al. 1990; Wong et al. 2003). In Nigeria, for instance, studies of clients showed that the reasons for nonuse of condoms were related to the level of intimacy shared within the relationship (Messersmith et al. 2000). In the Dominican Republic (Kerriga et al. 2003),[7] research on female sex workers has shown that the courtship-like process between sex workers and clients by which trust or affection are established over time leads to decreased condom use. In Zambia, women's desire to satisfy their sexual partners to retain their loyalty and faithfulness was an important reason for nonuse of condoms (Kapumba et al. 1991).[8]

The relationship between hostess Lin and her client lovers resembles marital couples. Hostess Lin was kept by her client as a mistress, and their sexual behaviors were conducted in the apartment rented by her client lover. Cohabiting together in the same apartment signified the closeness and intimacy of their relationships and hence influenced nonuse of condoms (see Cusick 1998).[9] Studies elsewhere have also found that sex exchange encounters at prostitute's homes were less likely to involve condom use because it indicated special relationships and arrangements between the clients and the women. A similar finding was also reported by Hansen, Lopez-Iftikhar, and Alegria (2002), in a qualitative study of Puerto Rican sex workers. Researchers have observed a close association between kissing and unprotected intercourse, which, again, pointed to the relationships between intimacy and condomless sex.[10]

ECONOMIC DEPENDENCE

Hostess Li was infected with an STD. She said, "I might have gotten it from Gao because we had condomless sex five times last month." I said, "Why didn't you use condoms or contraceptives?" She replied, "He assured me that he would use the withdrawal method and there shouldn't be a problem." I said, "You didn't insist in using condoms?" She replied, "He gave me 10,000 yuan and I took his money. When he did not use condoms, how could I resist? I took his money! Yes, I was afraid of diseases, but it was really hard to say no. He took me to a hotel five times last month and did not use condoms even once." Li said that she started noticing symptoms in her vagina after these sexual exchanges. She went to the hospital and was told that she had acquired a venereal disease. She said, "It cost me 2,000 yuan to finally cure it." I asked her, "Now that you know that it may be he who has given it to you, in retrospect, would you have requested or insisted on condom use?" She shook her head and sighed, "No, I don't think I would because, you know, I took his money. It's a lot of money—10,000 yuan. I spent 2,000 yuan on the disease, but I earned 8,000 yuan from him. It's still worth it."

Indeed, regular partners pose a potential threat for hostesses' health because condomless sex with regular clients account for a large share of all sex acts. As Li's case indicates, engaging in unprotected sex with client lovers led to unwanted pregnancies and infections of STDs. During the time of my research, most of the hostesses I worked with had undergone multiple abortions and STD infections by their client lovers. Each time when they contracted STDs from their clients, it cost them several thousand yuan to treat the diseases. When I asked why they did not use condoms to prevent STDs, their responses were invariably, "Clients didn't like them."

One of the hostesses I roomed with in the summer of 2005 had a rough abortion and was bedridden for almost a month. Her client lover paid for the abortion and visited her for several days. While lying in bed frail and fatigued, she said to me, "I suffer so much these days, and all these sufferings are for him." Hostesses wished that their sufferings for the men could move the clients and earn themselves more financial benefits. In this case, she was successful. Her client lover expressed appreciation for her sufferings and provided her with a large sum of money for compensation. She said, "I told 'glasses' (one of her client lovers) that I got pregnant. He immediately gave me several thousand yuan and accompanied me to the hospital to get an abortion. Ever since the abortion, he treated me even better and always called me 'wife.'" Apparently her abortion elicited guilt and responsibility from her client lover that made her seem more precious and more like a "wife" to him.

Hostess Lan also got pregnant by her client lover. He paid for her abortion and took a couple of days' leave from work to accompany her. I visited her a number of times with a couple of hostess friends. Lan and her client lover called each other "husband" (laogong) and "wife" (laopo). During our visit, Lan kept complaining, "I've suffered so much ever since the abortion." Chinese customs have very strict rules for post-abortion care. According to these cultural customs, Lan had to lie in bed for at least a month and adhere to the taboos on eating, drinking, and dress. For instance, although it was a hot summer, she had to wear thick clothes to preclude the damp and cold air from penetrating her body. She could not drink or eat anything cold for the same reason. She was sweating under bed sheets all day long, and she was not happy about it. Yet her sufferings did not come with nothing. Her client lover, feeling guilty and sympathetic for her, gave her thousands of *yuan* and bought her nutritious and strengthening food to help her recover quickly. Generally, abortion as a result of condomless sex earned hostesses more status and more care from men.

However, not all hostesses were so lucky. Some were simply abandoned by their clients after they became pregnant and had to "eat the bitterness" on their own. Therefore, it was always a gamble for the hostesses because they did not know how their client lovers would react until they got pregnant.

Yet hostesses were willing to compromise their health to attract and secure their client lovers—the source of their livelihood. Hostesses' health sacrifices for the sake of men reminded me of the Bulawayo women in Africa (Civic and Wilson 1996). The Bulawayo women, to satisfy men with dry sex, applied drying agents to their vagina, from which they suffered lower abdominal pain and internal infections. Other side effects included sores on female genitals, bruised skin, vaginal swelling, cuts, and abrasions. They were also reluctant to use condoms because condoms would block the "love potion" effects of the agents and stopped their magic. Despite these health sacrifices, women continued to use drying agents because of the positive effect on men's libido. They believed that the effects of drying agents attracted and kept a sexual partner (Civic and Wilson 1996).[11]

Why would hostesses sacrifice their health like the Bulawayo women and endure all sorts of physical problems? The answer was their economic dependency on men. Studies have pointed out that economic dependency involved in sexual relationships has a profound ramification on condom negotiation and condom use (Calves 1999). In addition to love, trust, and intimacy, economic and survival considerations are crucial to condomless sex (Oladosu et al. 2001).[12]

Hostess Min's words helped crystallize the crucial element of economic and survival considerations in condomless sex, "How can I negotiate condom use with the man who gives me money and takes care of my survival needs? You don't negotiate use of a barrier with the person who feeds you. You try your best to please him and make him happy. You are cautious not to anger him or upset him because it could jeopardize your livelihood. You try your best to create an intimate spouse relationship, even if it means your health is compromised."

Money and timing are two key factors in most hostesses' decision to not use condoms. When in desperate need for money, hostesses would disregard the risk of noncondom use and choose to take the risk. For instance, hostess Han's mother was hospitalized and called her to cover the hospitalization bill. It was an emergency because unless she paid it in full her mother would be evicted from the hospital. Under this circumstance, she called up one of her previous clients and stayed at a hotel with him for a whole night. The next morning she deposited 5,000 yuan, the money she earned from the client, to her bank account so that her mother could withdraw it on the same day. In this situation, there was no negotiation of condom use because it was considered trivial compared with her family emergency—her mother's desperate situation. In another situation, hostess Lin had not been chosen by clients for a couple of weeks and grew increasingly anxious each day. She had been draining her bank account, yet she needed to eat, pay her apartment rent, and send money home. One night, finally, a client came in and requested sex with her. She jumped at the offer, with much joy for the prospect of release from her financial pressure. Whether condoms would be used was not the focus of her concern. After she finished with the client, I asked her whether he had used a condom, and she replied, "No."

The case of hostesses resonates with studies of sex workers elsewhere (Wojcicki and Malala 2001). Many sex workers in southern Africa, for instance, insisted that they chose not to use condoms out of economic desperation. Studies of Baganda women also showed that economic needs constituted an important factor for nonuse of condoms (Mukasa et al. 1992). Indeed, if sexual partnerships involve financial gain or increased financial security, then it is difficult for them to request condoms because they recognize the potential economic harm. It is axiomatic that the economic context within which risky behaviors occur is critical (Rwabukwali et al. 1991; Schoepf 1988; Schumann et al. 1991).

Hostesses used abortion and STD infection as a bargaining tool to procure additional money from their client lovers. This pattern resonates with research conducted elsewhere, which also shows that nonuse of condoms can be a critical bargaining tool used by women to obtain extra money from their exchange partners (Wojcicki and Malala 2001).[13] Although their sexual relationships with clients vary in the amount of financial gains and the stability of relationships, the hostesses' goal of establishing a long-term, intimate, and romantic relationship with their clients is constant. The economic dependency involved in the sexual relationships between hostesses and their client lovers constitutes an invariable barrier to condom use.

HOSTESSES' PERCEPTIONS OF CONDOMS

In the survey question about "the reason for nonuse of condoms," 59 percent of the hostesses chose "trust," 29 percent of the hostesses responded with "the worry to convey mistrust," and the remaining 12 percent responded "it is too expensive." Regarding the question, "When do you initiate condom use?" 96 percent of the hostesses chose "with strangers" and only 4 percent chose "worried about STDs."

Hostesses' responses indicate that condom use is associated with infection and mistrust between strangers. Such a perception of condoms resonates with previous research that shows an attachment of condom use to prostitution and casual sex (Hart et al. 1999; Kaler 2004; Karim et al. 1992; Lamptey and Goodridge 1991; Lindan et al. 1991; Maharaj 2001; Marandu and Chamme 2004; Mehryar 1995; Mnyika et al. 1995; Taylor 1990b). In rural Tanzania, for instance, villages link condoms to promiscuity and infection and only use condoms with risky partners (Plummer et al. 2006). In the research on gay men in Norway (Middel 2001), a typical response is represented by gay man A, who, in the presence of condoms prepared by his partner B, assumes that B has been sexually active and has had many partners. In this situation, condom use makes A feel that he is only B's most recent partner in a long row of partners and not likely to be the last.

Similar to the gay men in Norway and villagers in Tanzania, to the hostesses, condom use is strongly related to a lack of trust and interpreted as offensive, suspicious, and suggestive of infidelity and mistrust. Hostesses do not want to jeopardize their relationships with regular clients, especially because asking for condom use may raise suspicions of infidelity. Hostess Shen said, "Requesting condom use is tantamount to telling him that I have slept with someone else and I am carrying diseases."

Other research has also shown similar results. Research in Cleveland, Ohio (Sobo 1995a), has shown that ethnic women described condomless sex with their sexual partners as "a sign of trust," "honesty," and "commitment." Using condoms announced to partners that they were not sexually exclusive and signaled a lack of mutual trust. Because of a strong association between condoms and extraconjugal sex, condom use only denoted failure in a relationship. Since safe sex would signal infidelity, women actively used unsafe sex to convince themselves that their men did not stray sexually. They did so as part of a psychosocial strategy for building and preserving an image of themselves and having achieved the conjugal ideal.

Research in Baltimore, Maryland, has also shown that a number of low-income African American women were reluctant to request condom use from their male partners (Wiutehead 1997). Their reluctance was based on their need or desire to maintain their ongoing relationships and the fear that men would leave them because of their strong dislike of condoms. Women were also very cautious about initiating condom use in their sexual practices because they feared that men would associate such practices with a woman being "sexually loose." If a woman initiated discussions of condoms, she might be suspected of promiscuity. Men interpreted this action as an insinuation that the women had been sleeping around and carrying diseases (Wiutehead 1997). Therefore, women were afraid to bring up condom use, fearing that it could threaten the survival of the relationship. Research in Greece has also revealed the same findings.[14]

Like the ethnic women in Ohio and women in Baltimore, Maryland, hostesses associated condom use with STDs, mistrust, and infidelity. Unlike the ethnic women who deliberately chose unsafe sex to achieve high self-esteem and a monogamous conjugal ideal, hostesses resembled the women in Baltimore in that they were afraid to initiate condom use, as it signaled their infidelity and that they were carrying STDs. As mentioned in the previous section, hostesses rely on their regular clients. To keep their relationships with regular clients, hostesses found it extremely difficult to propose condom use because they feared that the negative association of condoms would arouse clients' suspicions of them and end the relationships. Many of them chose to follow the clients if clients abjured condom use, for fear of losing them and, in turn, affecting their survival.

Similar to these studies, the case of hostesses reveals that an absence of condoms is perceived as essential in a love relationship. This notion of a love relationship is shaped in a cultural context where condomless sex is interpreted as a sign of love and trust, and safe sex is inexorably intertwined with deceit and disease. Hostesses' fear and anxiety of revealing their desire to have safe sex indicate the emotional nature of condoms (see also Middel 2001). Within a sexual relationship with regular clients, they felt it difficult to convey their desire for safe sex, fearing that condoms would signify mistrust, infidelity, or diseases. Therefore, some subordinated their fear and their wish to use condoms to show trust in their partner.

COPING WITH CONDOMLESS SEX

Previous research has argued that condom use is higher when women ask men to use condoms (Weissman et al. 1989) and when women believe that they can persuade their partners to use condoms (Kegeles et al. 1989). This study shows that in the hierarchical relationship between sex workers and clients some sex workers cannot express their desire to use condoms. Nor do they feel that they can talk clients into using condoms.

In Chapter 4, I related a story of how clients deliberately inflicted violence and pain on hostesses' bodies to exercise their power. In that story, hostess Lin turned down her client's invitation to attend a banquet and apologized to him properly. In revenge for her failure to meet his demands, the client took her to a hotel, penetrated her without a condom, and inserted a folded, brand-new 100 yuan into her vagina. The sharp and scratchy piece of paper expanded and exploded in her vagina, causing a hemorrhage. It took her several months to recuperate.

Studies elsewhere have shown that sex workers wield less control over condom use in unfamiliar or unregulated settings (Lau et al. 2003; Pyett and Warr 1997). In many cases, clients told hostesses to follow them, and they took hostesses to strange and unfamiliar settings such as hotels or empty apartments. In those places, hostesses usually could not have any say about condom use. In this case, hostess Lin's client took her to a hotel, forced her to have condomless sex, and inflicted traumatic violence on her body.

Scholars have argued that male violence and coercion with condomless sex is more prevalent in developing countries (Campbell et al. 2001). In the context of sexual coercion and paid sex, high-risk unprotected intercourse is more common (Macaluso et al. 2000). Clients' dislike of condoms, compounded with physical violence, thwarted women's negotiation of condom use or even refusal of sex. In my research, hostesses reported that they could only reduce their risk to a certain extent, since they could not alter male behavior.

SQUATTING

To reduce their risk, hostesses invented a repertoire of coping mechanisms against unwanted pregnancies and contraction of STDs. The first method was squatting. During my research at a karaoke bar, a hostess was taken to the upstairs sex room to engage in sex with a client. After about a half-hour, she came down to join us. Hostesses asked her how it went. She said it went OK. I asked, "Did you guys use condoms?" She shook her head and replied, "No." I asked, "Why not?" She said, "The client didn't want to use one." Hostesses were apparently very familiar with this situation, as no one appeared surprised. Rather, they immediately suggested, "Go to the restroom and squat over the toilet for five minutes. The sperm and bacteria will come out that way." The hostess followed this advice. Although seemingly a primitive idea, the squatting posture was considered useful and helpful coping mechanism through which the sperm and bacteria could be forced to flow out. Hostesses believed that squatting could protect them from getting pregnant and contracting diseases.

EMERGENCY PILLS

The second method was emergency birth control pills, used against unwanted pregnancies. Hostess Hong told me that she had been relying on emergency birth control pills because her client lover did not like to use condoms. Since she had been taking too many emergency pills, she complained to me that she had been suffering terribly from the severe side effects, such as heavy vomiting and irregular periods. Later on, she replaced emergency pills with regular contraceptives called Mafulong and took them daily and monthly.

Hostesses used emergency and regular contraceptive pills to prevent unwanted pregnancy with client lovers. How did they protect themselves from contracting STDs and HIV?

CHECKING AND WASHING

The third method was checking and washing. Hostess Hua said,

> My "husband" never uses condoms. I hate it, but what can I do? After all, he is paying for all my daily expenses, including the rent for my apartment. To make him use condoms, I bought condoms myself and deliberately left them in the most obvious place on the end table next to our bed. I was hoping that the sight of condoms could remind him to use them. When he wanted it [sex], I pointed to the end table. However, every time, he always said, "Wait a second, wait a second." And then he never ended up wearing them.

Hua has been a hostess for more than five years. I came to know her when I was studying hostesses in the karaoke bar Prince in 2000. Ever since then, we have been in close contact with one another. During these years, she has worked in different bars and has had countless client lovers. The interview was conducted in the summer of 2007.

As a key informant of mine, Hua was in charge of distributing my survey questionnaires to her hostess friends and ensuring that the surveys were answered carefully. After working with me for a while, she became aware of the importance of condom use and the risk of HIV infection. As she related to me, the knowledge of HIV made her extremely concerned about the condomless sex with her client lover. However, despite her wealth of knowledge, she was helpless because he had been paying for her daily expenses and supporting her livelihood. I asked Hua how she dealt with the situation that her client lover refused to use condoms. Hua replied, "I hate it! But what can I do? I have no way but to check his 'little brother' [penis] carefully to see if there are any obvious blisters, growth, or abnormal secretion before we have sex. After he is done, I immediately rush to the shower and clean my 'little sister' [vagina] carefully and thoroughly."

Since Hua's client lover did not use condoms consistently, Hua bought condoms herself and placed them at the most obvious place next to the bed. However, her efforts reaped no results. He continued condomless sex and ignored her implied request for condom use. Hua felt vulnerable and was worried about unwanted pregnancy and transmission of diseases. Yet she could not confront him. Nor could she threaten him because her subsistence was on the line.

Though frustrated, Hua's submission to her lover's unprotected sex reaped benefits. Her lover was responsible for the apartment rent, bought her a computer, paid for her computer classes and driving lessons, and hired her to work in his company. He had also promised to buy her a house in Shanghai, where he was from. She said, "I am going to buy myself a house in Dalian. That way I'll have two houses. I'll do business here in Dalian. When I'm old, I can spend my time either here in Dalian or there in Shanghai."

In many hostesses' eyes, she "had it made"—she had become successful. This success, as they were all aware, did not come without personal sacrifices. Among these sacrifices, relinquishing concern about her health was certainly one of them. However, it was a worthwhile sacrifice, as she rationalized to me, "After all, no pains, no gains. I earned this good life by forgoing concern about my body. It's worth it. Don't we all lose something before we can gain?"

Indeed, in the coterie of hostesses, unwanted pregnancy and contraction of STDs were so common that no one was surprised at others' sufferings. For instance, two of Hua's hostess friends had contracted genital warts from their client lovers. They went to the hospital and received a laser treatment to remove the growth. Each of them spent 300 yuan on the process and both were treated. Other friends of hers had contracted gonorrhea or vaginal lice. For the lice, they had to scrape off the hair, spray on the medicine, apply the lotion, and wash the area every day. Hua said, "Other than AIDS, every STD can be cured."

Hua had contracted STDs and suffered from unwanted pregnancies many times. She was afraid when her client lover would not use condoms. Yet she could not do anything because he provided her livelihood. To cope with the situation, she resorted to checking his genital area before sex for secretions, growths, or blisters. After sex, she rushed to the shower to clean her vagina thoroughly.

ANTIBIOTIC SHOTS

In addition to checking his genital area before sex and washing her vagina after sex, Hua sometimes took antibiotic shots before sex. I was curious about these antibiotic shots, so I went to several local hospitals to interview STD doctors. In a municipal specialized STD hospital, the director, Dr. Wang said, "We have a lot of hostesses coming here to receive the shot before they go out with their customers. My advice to them is that it has to be one to two hours prior to sex, otherwise it will lose its effect." I asked, "Can this shot protect the women from contracting STDs and HIV?" He replied, "This shot cannot prevent them from getting the disease. However, it can make the diseases that they have contracted curable."

I had never heard or read anything about the effect of presex antibiotic shots on procured STDs, nor had I known that hospitals actually provided these shots to sex workers. After my interview with Dr. Wang, I remained suspicious, as I was aware that medical research has proven that herpes and HIV/AIDS are both incurable. So I visited a local general hospital and interviewed the director of the STD department. The director evaded my question about the effects of antibiotic shots on STDs and HIV/AIDS, and said, "People believe that antibiotic shots work. You might not know this, but TV reporters who cover HIV patients always get transfusions of antibiotic fluids before eating and drinking with the patients. The idea is that, since you have the antibiotic protection, which boosts your immune system, you are less likely to be infected."

The director's words were confirmed by my interviews with a couple of journalists in Beijing. In Beijing, I met with journalists who had interviewed AIDS patients in Henan province—the province that was plagued by HIV infection because of the blood scandal. When they recounted their interview trip to Henan, they emphasized how scared and frightened they were about possible infections from their interviewees. They described in detail how they were terrified when they had to eat from the same dish and drink from the same bottle with the patients. To safeguard their protection, before setting off on the interview trip, they received transfusions of antibiotic fluids.

Like these journalists, hostesses resorted to antibiotic shots before sex to enhance their immune system and protect themselves from infections.

PILLS, HYGIENIC NAPKINS, AND LIQUID CONDOMS

In addition to STDs, hostesses were also subject to vaginal infections because customers always stuck their fingers inside hostesses' vaginas and kept poking their vaginas until they bled. As a result, even though hostesses relied on the previous strategies for protection, they had to take a lot of pills for vaginal infections. Hostess Cheng said,

> Customers always stick their finger into hostesses' vaginas and poke the insides. They will not stop until blood flows out. It hurts a great deal and causes a lot of infections. The result is that we always suffer from infections in the vagina and have to get them treated at the hospital. I always take pills for infections called "Women's Thousand Gold Tablet" [*fuke qianjin pian*] and "Golden Chicken Capsule" [*jinji jiaonang*]. I also use external washing liquid [*waiyong xiye*] to wash my vagina or hygienic napkins [*weisheng jin*] to wipe my vagina. Liquid condoms are also great because customers do not know you are using them.

As hostesses complained, customers' harsh and constant poking led to nonstop vaginal infections. To prevent infections and alleviate physical pain, hostesses resorted to oral pills, external washing liquid, hygienic napkins, and liquid condoms. External washing liquid was available at all drug stores and supermarkets. It came in over twenty different brands, such as Cleanse Your Vagina (*jie'er yin*), Treasure for Vagina (*Fuyin bao*), and Heal Itch (*keyang shu*). The package read as follows: "This product kills viruses and bacteria and prevents transmission of STDs. As a Chinese herbal medicine, it adheres to its principle of shedding the poison and ridding the toxin. It contains precious Chinese herbs such as cortex phellodendri [*huangbai*] and radix sophorae flavescentis [*kushen*]."

Hygienic napkins were also presented in numerous brands. The package of a napkin called Strong (*xiongwei*) read, "Private Cleaning Wet Napkin contains special ionic medicinal liquor. It has a highly effective and quick sterilization. It dispels bad smell and terminates itches. It prevents venereal diseases. It can also recover membrane and offer an ideal self-cleaning protection. It is easy to use, and there is no need for basin or water. It is a necessary health product to keep in the house or carry while traveling."

As advertised on the package, similar to external washing liquids, hygienic napkins were presented as capable of preventing transmission of STDs and sterilizing bacteria and viruses. Both items were advertised as protective means against venereal diseases. The hygienic napkins were used not only by hostesses but also by male customers. In my interviews, male customers believed that hygienic napkins were effective to kill viruses and bacteria and cleanse their genital organs. They especially emphasized the convenience when they engaged in sex in a place that did not have a water supply (e.g., in the car). Hygienic napkins became an indispensable necessity stored in their cars or their briefcases.

Aside from these two items, a liquid condom was another product that claimed to kill viruses and bacteria and prevent STDs and HIV/AIDS. Membrane-less liquid condoms were the most expensive among these three products. The package of the most famous brand *Kanglebao* read, "This product can kill all STD viruses and bacteria in one minute and the AIDS virus in 10 minutes. It has been tested and approved by the Science and Technology Research Center of the National Population and Family Planning Committee, National STD Control Center, Center for Disease Control, and Beijing University Medical Department."

The package of another brand read, "This product was co-invented by China and the U.S. It can annihilate all bacteria, germs, and sperms in the vagina. It can be used either before or after sexual intercourse to exercise these effects. Made from Chinese herbal medicine, it is mainly used to sterilize and kill all germs and bacteria in a vagina and penis. Just squeeze the medicinal liquid stored in the tube into the vagina."

This liquid condom was ubiquitous in drug stores and adult health shops, ranging in price from 20 yuan to 60 yuan for each tube. The liquid is squeezed from a syringe into the vagina. It was reported in 2002 that China had successfully produced an external ointment called "liquid condom" that could rapidly kill sperms, bacteria, and viruses, including AIDS, syphilis, and genital herpes (Zeng 2002). Liquid condom was commended as one of the 2002 National Torch Plan Fruit Projects by the National Science and Technology Department. Although liquid condom was tested and approved by national authority departments (M. Ren 2003), at least two manufacturers had been sued in court by a number of women who were pregnant after using them (Li 2005; Meng 2006). Despite the notorious lawsuits and that these two brands were proven to be ineffective in killing sperms, sales of liquid condoms were unaffected. Indeed, liquid condoms are still on the market.

Hostesses' secretive use of liquid condoms echoes sex workers in southern Africa who agree to condomless sex yet surreptitiously use female condoms (Wojcicki and Malala 2001). Sex workers elsewhere were also reported using vaginal microbicide gels to prevent HIV and STDs (Reed and Weinberg 1984).

Despite the questionable validity of liquid condoms, hostesses were avid users of them. Some hostesses in my research became pregnant or contracted STDs despite their use of liquid condoms. However, rather than attributing them to the failure of the product, they rationalized that it was because they had misused them.

I asked hostess Shan what kinds of contraceptives she had been using with her client lovers. She responded without hesitation, "Liquid condoms." She said, "In our bar, hostesses recommended this product to each other, and that was how I learned about it. Ever since then I have been using *Kanglebao* liquid condoms." The *Kanglebao* brand was the most famous brand, and it was as expensive as 60 yuan for each tube. Shan said, "It's very convenient to use it because men don't notice it. It's just excellent." However, months later she got pregnant from her client lover and had to have an abortion. I asked her, "What happened?" She responded, "I don't know. Maybe it's because I didn't use it a couple of times and got pregnant then." Since she had several client lovers, she could not even figure out which one impregnated her. Hostess Lan also went through the same ordeal—she used liquid condoms off and on and got pregnant with by client lover.

Just as hostesses were often not sure who impregnated them, at times they did not know who gave them STDs. This is not only because they had multiple client lovers but also because they did not have regular physical checkups. Hostesses usually either went to the hospital or purchased some medicines at drug stores or adult health shops if they noticed certain irregular or abnormal symptoms.

JUDGMENT

Both hostesses and clients relied on their own judgment to assess their partners. In the interviews, when I asked, "How do you know if someone carries STDs or HIV?" Both hostesses and clients answered, "First, you look. Second, you smell." They explained that "look" has dual meanings. First, you look at the person and see whether he or she looks clean and nice. For the clients, a woman is disease-free if she looks clean and pure. For the hostesses, a man is disease-free if he looks nice, has a respectable job, is educated, or holds high-ranking positions. Then, you look at their genital areas to check for any abnormalities. Before sex, hostess Hua always checked her client lover's penis for any erratic signs.

Hostess Huang contracted a STD from her client lover who was a director at a high-tech communications company. She said, "He's a very accomplished and successful man. He's educated and highly capable. I admire him so much that it never occurred to me that he could carry STDs." Similarly, client Jin told me that he picked up a "pure" hostess in the south of China during a business trip and had condomless sex with her for a week. At the end of his business trip, he brought her back to the north and kept her as his mistress. After a month, he found his genital area infected. After he saw the doctor and was told he had contracted an STD, he was so outraged that he immediately abandoned her. He repeatedly said to me, "She was so pure and so sweet that I never thought she could have an STD."

Research elsewhere has also shown that people rely on their intuition and cultural stereotypes to judge whether a person has STDs or HIV. People are ready to drop their guard in the absence of any intuitive red flags. Those red flags would have risen if a partner had evoked the person's stereotypes about who has STDs or HIV (see Carrillo 2002, 253). In South Africa, young men in Khutsong rely on appearance and reputation to decide whether certain women are safe and therefore not requiring condoms for sexual intercourse (Campbell et al. 2001). Men in Mexico were reported to trust a clean, respectable, nice, and healthy person and mistrust someone who is dirty, poor, and badly educated (Carrillo 2002, 248–49). People in Tanzania (Klein et al. 1999) refer to a person's general characteristics and behaviors such as church attendance and dress to determine whether a person deserves sexual trust (Longfield et al. 2002). Young people in South Africa also believe that they can filter out partners dangerous to their health through categories of "clean" or "unclean" based on their social interactions and appearance (Waldby et al. 1993).

Molm and his colleagues define trust as "expectations that an exchange partner will behave benignly, based on the attribution of positive dispositions and intentions to the partner in a situation of uncertainty and risk" (Molm et al. 2000, 1402). Hostesses and clients in my research also resort to judging one another's appearance, education, job, position, and personality to determine whether their partners are disease-free and worthy of trust. Confident about their judgments of one another, they tend to dismiss sexual history or STD/HIV status and erase risks involved in condomless sex. They feel safe from infection because they trust their own assessment.

CONCLUSION

The central and crucial role of clients in hostesses' life can never be underestimated (see Miller and Neaigus 2002). Hostesses purposefully keep regular clients to maintain their livelihood. Submission to clients' condomless sex constitutes one of the means through which they achieve this goal. Hostesses choose not to insist on condoms because their fear of poverty is greater than their fear of STDs and AIDS. In this sense, their subordination to clients' condomless sex is "an act of choice, as well as of dependence, that challenges conventional notions of agency and victimhood" (Wawer et al. 1996).

Researchers have argued that condom use is determined by a complex assessment of each relationship, the nature of the relationship, the level of trust in the relationship, the power and ability to assess and control sexual acts, the perceived risk of contracting an STD, and the future prospects of the relationship (Messersmith et al. 2000). In the case of hostesses, certainly they perceived the risk of STDs and certainly they feared HIV. Yet they exhibited little interest in learning about HIV because, to them, economic survival overrode health issues. Rather than health, their priority is maintaining regular clients, which they can tap into any time when they are in dire straits or need immediate financial help. Their goal is to establish a trusting, loving, and intimate, spousal relationship with regular clients in the present and in the future. As condomless sex is inextricably intertwined with love, trust, intimacy, and romance, it is essential for hostesses to engage in condomless sex with regular clients to create and strengthen a quasi-marital relationship. Compliance with client lovers in condomless sex helps convey their devotion and sacrifice, for which they expect financial compensation. As related in this chapter, hostesses endured repeated abortions; health problems, such as heavy vomiting and multiple periods per month as a result of emergency contraceptive pills; STDs infections; and HIV risks. Indeed, some of them have been rewarded with financial gains and a materially rich life through their sacrifice, which they deemed as successful and, hence, sufficiently worthwhile.

Insistence on condom use is perilous and ominous as it conveys infidelity, disease, and mistrust to their partners. When condoms are not an option to protect them against infections or pregnancy, hostesses rely on other venues such as presex shots, emergency contraceptives, liquid condoms, external-use liquids, hygienic napkins, and cultural judgments. Unfortunately, none of these recourses provide effective protection against STDs or HIV.

Despite the possible failures of these vehicles for protection, these micro-maneuvers constitute hostesses' "weapon of the weak" (Scott 1976). Campbell has also pointed to South African sex workers' self-empowering process in dealing with difficult environments, such as reworking the concept of respectability and networking among sex workers and in the squatter community. Wojcicki and Malala (2001) have also argued that we should not consider these women as victims only, as it dismisses their decision making, albeit at the micro-level.[15]

In her ethnographic study of female sex workers in a large British city, researcher Teela Sanders (2004) argues that sex workers construct a continuum of risks that prioritizes certain kinds of dangers. Although health-related matters are a real concern to many women because they generally have comprehensive strategies to manage health risks at work, this risk category is given a low priority compared with other risks. The risk of violence is considered a greater anxiety because of the prevalence of incidents in the sex work community. However, because of comprehensive screening and protection strategies to minimize violence, this type of harm is not given the same level of attention that emotional risks receive. Indeed, the emotional risks of selling sex and the chances of discovery by their family and boyfriends is prioritized in the hierarchy of harms, followed by the risk of violence from clients, and finally, health-related risks.[16]

As with sex workers in Britain, hostesses also use a continuum of risk to prioritize and understand different risks in their lives. Indeed, hostesses encountered a range of risks in their everyday lives, including criminalization, physical violence, exploitation by managers and owners, arrests, fines, and imprisonment (T. Zheng 2007). In contrast to sex workers in Britain, many hostesses' family and boyfriends were not only aware of their profession but also appreciative of their financial support (see T. Zheng

2009). Therefore, unlike British sex workers, hostesses prioritized financial risks over emotional risks. Their economic dependence on clients and intracompetition with other hostesses precipitated hostesses to establish long-term, intimate, and romantic relationship with their regular client lovers. Going condomless was one of the strategies to ensure such a relationship and the financial benefits that come with it.

Chapter 6

Sexual Matters and HIV Risks in Male Clients' Everyday Lives

Introduction

As INDICATED IN THE INTRODUCTORY CHAPTER, CULTURAL meanings of sexual behaviors are central and crucial to understanding sexual transmission of HIV (see Micollier 2004b; Parker 2001).[1] The previous chapters have explored a range of cultural factors, including culturally prescribed responsibilities for contraception, cultural understanding of HIV/AIDS, and perceptions of condoms, STDs, and HIV/AIDS. These cultural factors are crucial to an adequate comprehension of the social dimension of HIV/AIDS.

Chapter 4 delineates concepts of masculinity and perceptions of condoms by male sex consumers. Cultural factors such as the clients' perception of contraceptive use as the woman's responsibility and the peer pressure to present an image of "bravado" or "valor" by abjuring condoms contribute to high-risk behaviors.

This chapter continues this analytical line of inquiry by unraveling sexual meanings shared by male sex consumers. How do male clients define sex and sexual desires? What do sexual practices mean to male clients? In what kind of cultural contexts do sexual practices take place? How do they conceptualize sexual desires? How do they perceive the differences between sex with wives and sex with hostesses?

This chapter seeks to answer these questions and explicates how male consumers' ideas about sexuality are shaped by the cultural environment and how they interpret and make sense of their sexual identities, desires, and behaviors. More specifically, it examines the source of their sex education and their definitions of sex and control over sexual desires. Central to this discussion is the process through which they integrate their definition of sex and sexual desires with their sexual practices and ideologies. I argue that their conceptualization of sex and sexual desires as purely biological, natural, and uncontrollable leads to their belief in passionate and irrational sexual behaviors, realized through sex with hostesses.

Data in this chapter are drawn from open-ended interviews that explore male clients' sexual histories, sexual matters, and, in turn, the cultural context of HIV risk. Discussions regarding perceptions about the nature of sex, expectations about how sex should be conducted, and how and by whom it should be initiated help lay a solid foundation for culturally sensitive HIV intervention designs, a topic to be elaborated in the next chapter.

SEX AND SEX EDUCATION

While interviewing male sex consumers, I asked two questions, "What is your definition of sex?" and "What is the source of your sex education?" The answer to the first question was the same: sex is an animal instinct. The answer to the second question was, "I've never had sex education at schools."[2] Client Li elaborated his answer, "There's no need for sex education because sex is a natural instinct and you just know how to do it. Thousands of years ago, in peasant societies, no one had sex education, but everyone had children." Client Zhang also said, "Sex is a biological need. You grow more [sexual] needs as your health gets better and your kidney gets stronger [Chinese medicine believes that male potency derives from the kidney]. I used to have a big belly. Ever since I changed my diet and started exercising every morning, I lost my belly and started having more sexual needs. At night, my penis will erect on its own, without any consciousness. That's sexual need. Sex is acting upon biological needs."

Client Zhang's words represented the answers that I received. If men in my research overwhelmingly defined sex as a biological need, what about women? How would women answer this question?

I approached some clients' wives and several local women and asked the same question. Though in different forms, women's answers were similar to one another. Some said, "Sex is the culmination of love." Others responded, "Sex is the ultimate expression of love."

It is axiomatic that the divergent answers from men and women in my research indicate that women link love with sex, whereas male clients' definition of sex is devoid of love. To further illustrate this gender difference in conceptualizations of sex, I would like to quote client Jin, who told me the following story:

> I have two friends, a man and a woman, both married. For years the man had desired the woman, and finally proposed sex to her. The woman declined. The woman said, "If I liked you a lot, I would do it with you. But I can't because I don't like you enough." For the man, however, sex is not so complicated. It doesn't mean so much. It just means a sudden biological urge or impulse [*yishi chongdong*]. It means the desire to possess [*zhanyou de yuwang*]. It has nothing to do with likes or dislikes. He can do it without such concerns.

Client Jin related this story of his two friends to highlight how men and women perceive sex differently. As he contended, to men, sex is a biological instinct, an urge, or an impulse. The biological nature of the sexual need means that its fulfillment has nothing to do with love. In the story that he narrated, the woman refused sex because her weak feelings for him did not justify sex. However, to the man, feelings were irrelevant. Sex to him was a purely physical thing—the desire to possess a female body and the desire to satisfy a biological impulse.

Sex as a biological instinct was also defined as exclusively heterosexual. During my interviews, it was ubiquitous that clients held heterosexual sex as normal and natural. Client Ping said, "The society should ostracize homosexuals as abnormal and unnatural. A normal man will never be gay. Only abnormal and sick men are gay. It may be because they have gone astray, or they are afraid of women, or they have undergone bad experiences with women, or they have had sexual abuses during their childhood." Learning that I was a volunteer in a local gay nongovernmental organization, Ping looked at me as if I were an alien from another planet. He said, "Don't tell others you are with them, otherwise you will be ostracized. How the hell can these guys be gay? Why would anyone give them funding? The country should outlaw them. The country should attack and imprison them."

Clients' sexual beliefs reflect their internalization of culturally prescribed sex roles. For them, learning sex is learning cultural norms of sex roles. Their definition and interpretations about sex as heteronormative and purely biological indicate their acquisition of social norms concerning sex roles and the management of sexuality. This social norm defines sex for men as instinctive and biological and defines sex for women as embodied by emotional commitment. In China, girls are inculcated with moral values about the perils of sex and the need to wait until later in life, preferably until marriage, to be sexually active. This cultural norm is a cultural effort to protect women's virginity and social order as a whole (Hershatter and Honig 1988).

Men in Mexico and South Africa manifest similar beliefs about sex. For instance, men in Mexico described themselves as hunters by nature—"A man could be sexual with little consequences" (Carrillo 2002, 46). Cultural norms in Mexico correspond with those in China not only in the tacit specification of sex roles but also in the culturally accepted and coercive heteronormativity. Like clients in my research, men in Mexico have internalized heteronormative sexuality, which allows them to fit into this moral core and not diverge from it (Carrillo 2002).

Similarly, in South Africa, social constructions of masculinity promote the norm that men need sex and women should refuse sex outside of committed relationships (Campbell et al. 2001; Moore and Rosenthal 1992; Ramazanoglu and Holland 1993; Wilton et al. 1991). In South Africa, society classifies "normal" men as having multiple partners and power over women. Young men are expected to adopt this masculine sexuality and overcome emotional vulnerabilities to be accepted as masculine in their society (Holland 1994).

In China, cultural norms recognize men's biological need to engage in sex without considerations of feelings or love. Popular media recommend that women forgive their husbands' extramarital affairs because men's biological nature determines their involvement in extramarital sex (see also T. Zheng 2006). As an article in a popular magazine *Hers* reads (Lan 2006), "All men love play. It is their nature. The reason that men can separate love from sex is because they are like animals. They emphasize sensory stimulation, not feelings. They have to depend upon the most basic and the most intimate contact to feel complete release. Men's macho role depends upon their conquest of women. Why does a man pursue pretty women [*meimei*] one after another? It is because he wants to prove to everyone that he is capable and he is a man."

Culturally normative sex roles explain men's sexual habits in biological terms and define men's sexual desires as detached from love and feelings. Indeed, as noted in Chapter 3, popular media are suffused with the gender norms of preserving male comfort and accommodating masculine sexual urges. As Linda Gordon contends, "It is easier and more 'normal' for men to be lustful and assertive, for women merely to surrender, to be carried away by a greater force" (Gordon 1979, 126).[3] Entrenched in such a cultural environment, it is not surprising that clients in my research have internalized these cultural values and reiterated biological determinism as immutable and embodying truthful natural laws. As shown in the next section, clients' cultural definition of sex lends its support to their interpretation of sexual desires as uncontrollable.

UNCONTROLLABLE SEXUAL DESIRE

During the interviews, I asked, "Should sexual desires be restrained?" The typical answer I received was, "Because sex is natural and biological, and all men think with their lower parts [penis], we cannot, and should not, restrain our sexual desire." I asked, "What kind of women can instigate this kind of desire?" One client said, "A woman with long hair, slender, pretty, and tall. Her clothing and adornment are in tune with my taste. Looking at her face and body, or feeling her body, or listening to provocative words makes me have sexual desires and wish to make love." Another client said, "The moment you see her, you want to possess [bazhan—monopolize] her. Your penis immediately gets erected. You feel the passion inside of you. When I have desires, I don't restrict them. I have sex, and I act according to my desires. Why restrict desires? I'm still young. I have passion. I need to release my passion." He continued to rationalize it with biological determinism:

> Our biology makes our sexual organs erect, and afterwards, we have to ejaculate and release it. Otherwise we would feel blocked [du de huang]. Let me give you an example. Some guys love to masturbate and they even do it in public restrooms, or do it many times a day. They cannot control themselves. Some guys may have just finished masturbating in the public bathroom, they walk out, see a pretty woman, and they have to return to the bathroom and masturbate again. It's because their desire is uncontrollable. Nowadays there is a sexual revolution. We can do anything we want.

"What if a woman that a man has desire for carries the HIV virus? Would you still say that he cannot and should not control his sexual desire?" I deliberately asked this difficult question to elicit his further reactions. He responded,

> Men don't care at the [heat of the] moment. Driven by his sexual desire, he doesn't care about anything but sex. Nothing could stop him because he is burning with sexual desire. Desire is a frightening thing. It's like a fire that can burn away a whole prairie [xingxing zhihuo, keyi liaoyuan]. When it hits you, you lose your mind and you can't control yourself. All you can think of is sex, making love, and releasing your desire [fangzong qingyu]. So you throw everything outside of your mind—you say, "I don't mind

even if I die afterwards." After sex, regret might cross your mind, but at the moment, it doesn't matter. It escapes your mind.

His words were confirmed by other clients' interviews. Client Li said, "We cannot think about contraceptives. It makes us distracted [*fenxin*]. We'll lose our attention and become impotent [*buxingle*]. It's like talking. Talking is like a stream of running water [*liushui*]. If it's suddenly interrupted and stopped, how disappointing it is [*duo saoxing*]. We would immediately feel disinterested. So I have never used contraceptives in my life. Women are responsible for getting IUDs."

"What about your first sexual experience? Did you use contraceptives?" I asked.

"No, I didn't," he answered, "because I had a sexual impulse [*xing chongdong*]. I simply couldn't control it. She had emergency pills afterwards. Later on we continued engaging in sex without contraceptives, because sex is spontaneous, and sexual desire is uncontrollable. You just don't prepare for sex. You have it when you feel like it—anywhere, any time. So you just don't carry condoms."

"What if she was pregnant?" I asked.

Li responded, "She did get pregnant. When pregnancy hit, we resorted to abortions. Currently a large portion of teenagers and college students have gone through abortions. Hospitals are full of these young people. We did it too before marriage. After five years, we got married. Ever since then, she has had an IUD for contraception."

Women should be responsible for contraception, as client Li and others contended, because men's sexual impulse is uncontrollable and sex is spontaneous. Rendering women responsible for contraception is not at all surprising. Indeed, it resonates with the analysis in Chapter 1 in which I demonstrated how the cultural environment demanded that women bear the brunt of family planning. This same cultural pattern is also found in many other parts of the world, such as rural Lebanon and Tanzania (Ku 2004; Plummer 2006). Paxson writes, "In these modern patriarchal contexts, the most fully feminine form of birth control is to manage the consequences of succumbing to sexual appetites . . . by dissuading a male partner from trying anything . . . or by being prepared to have an abortion or to birth a child she may or may not be able to raise properly" (Paxson 2002, 320).

The rationale clients offered is that men's sexual desire is spontaneous and overpowering. When sexual desire takes over (*jiqing yi shanglai*), it grips a man and makes his brain stop functioning. It makes a man indulge in the heat of the moment and unable to control himself. It strikes a man with such a force that it is "like a fire that can burn away a whole prairie [*xingxing zhihuo, keyi liaoyuan*]." It cannot be stopped or interrupted as concerns about contraceptives or diseases can make a man "distracted," hence he loses his erection and becomes impotent. Indeed, a man's sexual desire is considered overwhelming and compelling.

Clients' interviews remind us of research findings in Mexico. Similar to clients in my research, men in Mexico expressed sexual passion as an uncontrollable force (Carrillo 2002, 194). However, although they believed in the pursuit of sexual passion, men in Mexico were also convinced that sexual passion was dangerous, thus should be controlled. They perceived it imperative to strike a delicate balance during sex to allow themselves to flow spontaneously for true ecstasy but also to control the aspects of passion that may be destructive or threatening (Carrillo 2002, 195).

Like men in Mexico, clients in my research admitted that sexual passion is pernicious to society. Unlike men in Mexico, clients argued that because sexual passion is natural, it should not be regulated or controlled. Client Tin said,

> Releasing sexual desire indiscriminately could do damage to the society. When we were in school, we were taught traditional values to follow the dictates of heaven and extinguish human desire [*cun tianli, mian renyu*]. In the current new era, although the country can no longer implement this value system, they still teach us to curb and constrain our sexual desires. Because if we don't, it'll be like: "moving one hair, and the whole body is affected" [*qian yigen toufa, dong quanshen*]. That is to say, it would have a profound impact upon the whole society. It'll affect social morality [*daode*], ethics [*lunli*], laws, and the entire social system. It can even overthrow [*dianfu*] the whole political system.

Client Tin associated sexual desire with the stability of the political system and declared that sexual passion potentially threatens the polity. As he asserted, the state taught everyone at school how to reject their basic human desires. Despite the state's efforts to regulate and monitor everyone's sexual morality, as he claimed, the state had lost its currency among people. People such as him were no longer victims to this kind of teaching.

Clients targeted their criticisms directly at the Chinese Communist Party and the socialist polity (T. Zheng 2006). They considered Chinese Socialism and Confucianism as two parts of the same whole: an oppressive regulatory regime that stifles "human nature" and goes against the "natural way." Disillusioned, they turned to the hedonistic pursuit of pleasure and material wealth to compensate emptiness and meaninglessness. They regarded pursuit of sexual desires as a form of resistance against the artificial shackles placed on human sexuality by an unnatural social system.

As illustrated, clients recognized the dual nature of sexual passion as both destructive and fulfilling. Unlike men in Mexico, clients in my research specifically linked sexual passion with the socialist polity. They believed that the state vilified sexual desire because of its potential ability to overthrow the polity and threaten the life of the state. Resisting against the state's preaching that sexual passion should be circumvented and constrained, clients desired having a natural body filled with natural instinct that was free from the unnatural socialist system. Pursuit of sexual desire not only conveyed their disbelief and defiance against the socialist system but also served as a replacement for their inner moral vacuum (T. Zheng 2006).

WIVES

Clients commented in the interviews that their wives were not sexually open (*kaifang*), yet that was what they expected. Client Tan said, "If you ask my wife these questions, she won't be able to answer because she is too conservative." I asked, "Has your wife ever taken the initiative [in sex]?" He answered, "No, the husband is the one who should take the initiative [in sex], including postures. A wife shouldn't. This should be the case. Women should be more passive, and man should be more active. A wife shouldn't take the initiative because that would make me think she's the seductive kind of woman. Maybe after we're married for a decade and I feel I have already known her, if she initiates [sex] once, I might not conceive of her as the seductive kind."

Like Tan, clients agreed that wives should be conservative and not seductive, and the hallmark between the two is determined by whether the woman initiates sex. Women who initiate are considered impure, seductive, with high sex drives. Women who do not initiate are perceived as pure and proper. It is these women that clients desire as wives.

Client Gan said, "My wife represses her desires and feelings. She always clenches her teeth not to moan or show any facial expressions during sex." Other clients also confirmed his words and stated that their wives were cautious not to verbally convey what they wanted and how they felt about sex.

"What about hostesses or lovers outside of the marriage?" I asked. Client Tan said, "She should take the initiative. She should also propose and change postures so it'll be more enjoyable for both of us. Some women are most crazy [*fengkuang*]. The crazier they are, the better it [sex] is. This kind of woman says flirting [*tiaoqing*] words to me. For instance, she says, 'Come on' and hurries me to have sex with her. She touches me. She pursues stimulation. This kind of woman—how stimulating [*ciji*] and how exciting!"

Interviews revealed that clients harbored a clear double standard for a wife and a lover. They favored an open, passionate, and crazy extramarital lover but could not tolerate a wife who resembled that. The same finding was also found in Mexico and Japan where men envisages a wife to be conservative and a lover to be passionate. In Mexico, men described a wife as pure, decent, chaste, and conservative, whom they would not try different sexual positions or enact their fantasies with (Carrillo 2002, 43–45). Men also felt that the privilege of initiating sex was exclusively theirs and never the wives'. Both men and women perceived women who initiated sex as "loose" and believed that they should be careful about how they presented themselves to men to avoid being automatically placed in the "bad women" category. Men relied on the assessment of a woman's negative reaction to his incremental sexual advances to determine her as virginal and, therefore, a potential future wife. Therefore, women with free sexuality pretended to be submissive and inexperienced to be accepted as a more stable, potentially marriageable sexual partner (Carrillo 2002, 48, 184–85). In Japan, men were also socialized to perceive wives in the marriage as conservative, desexualized and de-eroticized, and women outside the marriage as objects of male desire (H. S. Lin 2008, 46).

To understand clients' wives better, I conducted interviews with them. My interview with Min represents the wives' point of view:

Q: How do you define sex?
A: Uh, sex is a means of reproduction.
Q: You've already had a kid, so are you saying that there is no point having sex any more?
A: Well, sex is not an entertainment. Sex is not for pleasure. Having sex with my husband is important, not for pleasure, but for the purpose of health. There are two kinds of sex—one is evil sex [*xie xing*] with someone who is not your spouse; the other is righteous sex [*zheng xing*] with your husband. The first one saps your energy and makes you weak. The second one keeps the spouses healthy.

Q: When your husband goes on a business trip and you have sexual desire, what would you do?

A: I'll divert [*zhuanyi*] my attention. When I feel the desire, I know that it's because my period is coming. The desire comes from my uterus that is replete with blood. I'll immediately divert my attention by doing something else. It's very easy to forget the desire and focus on other things. It's different from repression though. Repression means that although you repress the physical desire, you still suffer from the mental desire. That is, you can't stop thinking about it [sex] in your mind. That's painful and wastes a lot of energy. I'm different from that. I simply stop thinking about that.

Q: What would you do if you learned that your husband has had extra-marital sex?

A: It's not my problem if he does it. It's his problem, not mine. It harms his health, not mine. He gets tired easily. He was in poor health when he was little. So he's very careful about his health.

Different from clients who defined sex as a biological instinct, Min defined it as a means of reproduction and health. Min believed that sexual desires should be repressed, diverted, and sublimated, to focus on something else in life. She castigated people who wallowed in sexual pleasures. Indeed, she did not believe in sexual pleasure: "Sex is not an entertainment. Sex is not for pleasure." Sex, she declared, was for health and reproduction. Min classified two categories of sex: evil sex with an improper mate and righteous sex with a proper spouse. She contended that evil sex destroys a person's health, and righteous sex enhances a person's health.

Min's belief was reminiscent of traditional Taoist ideology about sex (Furth 1994). Taoist ideology promoted healthy sexual norms that were not about pleasure but were about life and death, longevity and posterity. The goal was to attain self-discipline and self-denial to transcend the body's ordinary limitations, defeating the entropy of age to achieve longevity and spiritual regeneration (Furth 1994, 137).[4] Women were advised to redirect their sexuality toward motherhood (Furth 1994). Clients' wives like Min in my research repressed and denied their sexual desires to safeguard the marriage and the family.

The Chinese government strives to ensure that sexual conduct, particularly women's sexual conduct, is confined to marriage (Hershatter and Honig 1988; Sigley 2001). To prevent sex from occurring outside the bounds of matrimony, the state has instructed young women to manage and to control men's sexual advances before marriage, satisfy their husbands' sexual needs, and stay clear of extramarital affairs after marriage (Hershatter and Honig 1988).[5] In their study of advice literature, Emily

Honig and Gail Hershatter note that it invariably rests on the shoulders of young women to "channel and control the sexual desires of young men as well as their own and to defer acting on those desires until they [have] reach[ed] the socially appropriate age for courtship and marriage" (Hershatter and Honig 1988, 53).[6]

Similarly, in South Africa, young women preserve their reputation by allowing men to initiate sexual encounters (Campbell et al. 2001). Greek women have also been "trained to expect and resist men's advances unless it is socially appropriate for relations to occur, in which case they are to be seen to submit to male desire" (Paxson 2002, 317). As Paxson writes, "Although the gendered ethics of appropriate sexual behavior dictate that women must demonstrate greater control because their moral character is weaker (in theory) and hence they are in greater need of restraint, in demonstrating such restraint they actually prove themselves stronger (in practice) than men, who can't help themselves in any event. This ideological contradiction has worked out well for the men, as women are left to shoulder (behind the scenes, as with abortion) the burden of procreative accountability" (Paxson 2002, 322).[7]

Sex with Wives versus Sex with Hostesses

Eroded Interest in Wives

In the interviews, clients invariably claimed that they had sexual desires for women who were not their wives. Client Lin said, "Sometimes you just want to make love to women who are not your wives." "Why?" I asked. "A wife is never perfect, you know," he replied, "She always has certain defects. When you see other women who don't have these defects, you start having desire for them."

According to the clients, their desire for other women was not only because these women could satisfy them in ways that their wives could not but also because it was inevitable that men eventually would lose sexual passion for their wives. Client Hang said, "Any couple, after being together for a while, grows tired of each other and loses passion [*jiqing*]. Every man knows that. After a couple of years, there is nothing new [*xinying de*] [in the relationship] any more. So you get bored and look outside for passion [*jiqing*]."

Because of the boredom, clients claimed that they only engaged in sex with their wives once or twice a month. Although they had lost interest and desire in their wives, as they contended, it was still their obligation (*yiwu*) to copulate with their wives once in a while. Client Mi said, "Each

time [with my wife] is a perfunctory act [*fuyan liaoshi*]. It's pure obligation [*yiwu*]. You just repeat the same old stuff and follow the routine
[*lixing gongshi*] with her. But it's different with hostesses. They give you
a sense of freshness [*xinxian gan*]. You feel more passionate and more
stimulated. You know, once two people are too familiar with each other,
the passion is lost. Only with strangers can passion be developed. You get
so excited [*hen xingfen*]. You feel so electrified [*laidian*]. It's like touching
electricity. It's extremely stimulating."

Client Mi's words testified to that sex with wives felt different from
sex with hostesses because familiarity bred contempt and strangeness
spawned passion. Like Mi, other clients described sex with their wives
as boring (*fawei*) and tiring (*pibei*). "Touching my wife's left hand is like
touching my right hand. I don't feel a thing. It's '*shenmei pilao*.'" *Shenmei
pilao* in this context means that, even if the wife is a beauty, the husband
gets tired of her because of the familiarity. "There's no passion, love, or
romance between my wife and me after marriage," client Hai said. "What
remains between us is relative-love or family-love [*qinqing*], like sisters
and brothers. Without any passion, I'm just following the routine and
completing the task [of having sex]. With a lover, however, there's a fresh
feeling."

Clients described marriage as a "besieged city" (*weicheng*) that suffo-
cated passion and represented "the tomb of love" (*aiqing de fenmu*). They
needed to "breathe some fresh air and smell some flowers outside." "A
wife is never able to be like a lover (*qingren*)," client Lin said. "Marriage,
love, and sex are three different things. If not, why did we invent three
words to call them?"

Having lost interest in their wives, clients asserted that the only thing
that held (*weixi*) the marriage together was the child. Client Shi employed
the metaphors of cars and houses to describe the difference between sex
with wives and sex with hostesses. He said, "It's like driving a car. If you
always drive the same car, you feel dull and uninteresting. After some
time, you want to change it to a new car. You feel stimulated and fresh
in the new car. It's also like . . . If you always live in the same house, you
feel bored. You want a change . . . After having a child, the child delivers a
fresh feeling, but between the spouses, it becomes dull [*pingdan*]."

While sex with wives was depicted as feeling-less and boring, sex with
hostesses was portrayed as "exciting and stimulating." Client Liu said,
"Transgression [*chugui*] gives you a great deal of excitement and stimulation [*xingfen he ciji*]. Sexual transgression is like playing truant in your
work. Furtively slipping out of your working unit to play is a lot more

stimulating and exciting than playing after work. It's because it's against the rules and common regulations. It gives you more excitement. It's much more stimulating."

Liu, a government official working at a municipal government bureau, found excitement in rejecting norms, whether in sex or at work. He envisioned the transgressive act of slipping out of the government apparatus to play as exciting as engaging in illicit sex with hostesses. As he contended, play after work was tantamount to sex with the wife. Both acts were obeying the state-prescribed rules and regulations and therefore boring and dull.

Clients compared sex with their wives as peasants turning in the grain tax to the state. The clients/peasants perceived themselves at the bottom of the hierarchy vis-à-vis the wives/state. Clients' "misappropriation" of their semen was a mode of resistance, just as grain was misappropriated by peasants who rebelled by cheating the government of their taxes. The clients' subversive misappropriation was intended to maintain their bodies' independence (T. Zheng 2009). In this interview, Liu compared sex with hostesses to playing truant at government mechanisms, embodying his defiance and resistance of the state-prescribed system and order.

Research of men in Mexico has also revealed that men favored the excitement of sexual transgression (Carrillo 2002). However, in this context, sexual transgression was thrilling because of the background of sexual silence in Mexico—the prevalence of a social environment not conducive to open, formal communication about sexuality among adults or sexual partners (Carrillo 2002, 149). It was sexual silence that contributed to an excitement associated with transgression, as it "flavors cultural scripts about seduction, sexual passion, and the enactment of sex itself" (Carrillo 2002, 149).[8]

Different from men in Mexico whose excitement derived from a social environment of sexual silence, clients' excitement stemmed from their rebellion against state-endorsed sex within the confines of marriage. As they contended in the interviews, their revolt against marital sex as the only legitimate sex was parallel to their defiance against the repressive state apparatus.

SEX WITH HOSTESSES FOR PASSION

Client Wen recounted a story in the interview that represented the differences between sex with wives and sex with hostesses. He said, "I don't have any passion for my wife. I don't like having sex with her. When I try to fulfill my obligation [of having sex with her], it only lasts about two minutes. I want to do it with hostesses because I can do it better and I enjoy it much more." He rolled up his shirt and revealed some red marks on his shoulder to his friends and me, "Take a look at this. Yesterday the hostess scratched me with her fingernails because the pleasure was too much for her. I was so excited that I did it for half an hour." Wen repeatedly commented that he enjoyed it so much that he became addicted to it. He admitted that he wanted to do it all the time with her.

Client Wen distinguished his obligatory sex with his wife from his passionate sex with his hostess. While the obligatory sex lasted only two minutes, the passionate sex lasted half an hour. He was also thrilled about the red marks his hostess had left on his shoulder. To him, they were the solid proof of fervent sex. As he described, she was so overwhelmed by the pleasure he gave her that she could not help but scratch open his flesh on his shoulder with her fingernails. The marks, of course, also became the hard evidence of his virility and sexual potency.

Clients in my research resonate with men in Mexico who do not regard love as necessary to justify sex and consider the search for pleasure and sexual passion to be good enough reason to have sex (Carrillo 2002, 187). Men in Mexico constantly talk about sexual passion and assign it a central role in their assessments of sexual interactions (Carrillo 2002, 193). For them, it reflects strong mutual attraction, compatibility between partners, and high degrees of sexual satisfaction, and its absence or opposite is sexual dissatisfaction or bad sex (Carrillo 2002, 193).

Like men in Mexico, clients opposed plain, or banal, sex with their wives to passionate sex outside of marriage, and asserted that sex outside of marriage invigorated and regenerated their masculinity. In their own words, "The crazier [fengkuang] a woman is, the more I love to have sex with her." "Sex should be a passionate experience [jiqing tiyan]. What we play with is heartbeat [wan'er de jiushi xintiao]." Hostesses, to them, could deliver the passionate experience of "heartbeat."

SEX WITH HOSTESSES FOR RELEASE OF PRESSURE

Clients not only depicted sex with hostesses as exciting and stimulating but also described it as a release of pressure. What precipitated this pressure? What did this pressure consist of?

In the interviews, clients pinpointed the source of their pressure as the desires of the society and their wives to become wealthy and surpass others in status. Client Ray said,

Because we live in a rich neighborhood, my wife thinks everyone is rich. She thinks that if it's easy for others to earn money, it should be easy for me too. I earned a million *yuan*, but she wants a billion. She complained, "Other men can make a billion or ten billions, why can't you?" She's unsatisfied with me. I said, "If you aren't satisfied and you think I cannot bring you happiness, I can return freedom to you." Then she became silent. Because of the pressure from my wife and the society, I seek release of my pressure in the sauna bar. You know, every day we are dressed up and we constantly compare with each other. [It is the comparison] that gives us pressure. When I'm at the sauna bar, everyone takes off their clothes and no one can tell each other apart. No one knows who is better or worse. Everyone's the same after the clothes are off. There's no difference. I feel very released, especially after having my body massaged and washed [*cuozao*], having my feet massaged, and getting a good sleep at a rented room, with a hostess. After coming out [of the sauna bar], I feel rejuvenated and energized [*jingshen dousou*], relaxed and light-hearted [*qingsong*].

For client Ray, the pressure came from his wife and the society. Feeling exhausted and tired of the pressure to exceed others in wealth, he found refuge in a sauna bar, removed from the society and his wife. As he described, the sauna bar was like an uncivilized world where no one wore clothes and social differences were leveled out. In such an arena where no one is different and everyone is equal, he felt a release of pressure. To him, body massage and hostesses' services not only enhanced the release but also "rejuvenated and energized" him.[9]

Clients identified another source of pressure as China's rapid social transformation. Client Jin, who was in his late thirties, said,

I grew up in the 1980s. My generation went through the most drastic social transformation. Since my generation, free college education was suddenly terminated. We had to pay for our own education [*zifei*]. Prior to my generation, insurance was never individuals' responsibility. It was taken care of by the government. My generation has had to buy our own insurance—medical insurance, life insurance, and pension insurance. Prior to my generation, apartments and jobs were all assigned free of charge by schools and working units. Nobody needed to hunt for jobs or apartments or pay bribes. There was no need to worry about these finances. My generation has had to buy our own apartment and find our own jobs. We had to shoulder all these responsibilities. Although the current young generation faces the same situation, at least they have already accepted it because these systems are already established and there is no challenge to them. We, however, had to adjust ourselves to these major and sudden changes and shoulder enormous pressure. We had to feel our way [*moda gunpa*] through the drastic social transformations and learn to accept these abrupt changes that have had a huge impact upon our lives. Facing such social pressure, I need to release it. Hostesses are a psychological therapy to me. I feel healed after engaging in sex with them. I can then rejoin the society with more energy and more confidence to face the pressure and the challenge. When I get beaten and feel down again, I go back to the hostesses, get recharged, and then return to the banality and pressure of my life. Such is the cycle of a man's life, between banality, pressure, excitement, and healing.

Client Jin portrayed the pressure as spawning from the drastic social transformation. Indeed, his generation bore the brunt of a sudden social change whereby pressure was heaped on them. Before his generation, college education, insurance, apartments, and jobs were all dispensed to Chinese citizens for free. Since his generation, none of these are free. He had to shoulder these finances by paying for everything, including briberies and strenuous efforts to secure a job. Of course, the current generation faced the same situation, but they did not expect a different scenario. Neither did they experience a sudden social change. Jin, however, had to readapt to the society. Sex with hostesses was indispensable in the process. It served as a rescue, a refuge, or a battery charger that released his pressure but also rejuvenated him to take up the challenge in his life.

CONCLUSION

This chapter has illustrated clients' own interpretations of sex, sexual desires, and sexual practices. Their interpretations are informed by the cultural assignments of gender and sex roles and, at the same time, opposed to the cultural enforcement of sex within the confines of marriage. On the one hand, clients desired to conform to social norms of heteronormative sex and proper sex roles. They expected a wife to be conservative, passive, and chaste; yet, at the same time, they longed for an extramarital hostess who was passionate, seductive, and stimulating to fulfill their fantasies and release pressures in life. On the other hand, clients rejected social norms that restricted sex only within conjugal limits. To them, regulating sex within matrimony was repressive as it clashed with men's biological and instinctive sexual needs. They deemed transgression of this social norm as exciting and stimulating, emblematic of their rebellion and defiance against state-prescribed moral codes.

Sex education was almost completely absent in their schools. Through their sexual explorations, clients defined sex as a biological and heterosexual urge and deemed sexual desire as uncontainable and irrational. Clients' sexual beliefs crystallized their internalization of the culturally normative sex roles. This heterosexual normative identifies men's biological need to engage in sex without considerations of feelings or love and recognizes women's emotional need to link sex with commitment.

In stark contrast with clients, women defined sex as a social practice that could only be acted out or performed upon interest in the man. Clients' wives ascertained that sex was not for pleasure or entertainment. Rather, it was for reproduction and health. Wives disdained the pursuit of sexual pleasure and strived to channel their sexual desires to work and motherhood to maintain the sanctuary of matrimony and family.

Clients expected their wives to be conservative and pure but longed for lovers or hostesses who were passionate and exciting. This criterion, coupled with what they deemed as "familiarity bred contempt," resulted in a "dull" and "boring" sex life with their wives. Sex with hostesses compensated for that boredom and brought them a thrilling and exciting experience.

Sex with hostesses was considered as transgressive and therapeutic. Clients associated the illicit sex and release of sexual desires with the current polity. Recognizing release of sexual desire as a threat to the polity, clients disregarded the state's insistence that they circumscribe or even reject their sexual desires. Instead, they insisted on pursuit of a liberated and natural body free from the constraints of the social and political system. To them, transgressive and illicit sex with hostesses embodied their resistance to the political order. Sex with hostesses was also therapeutic, as it provided a refuge for clients who felt trapped in a strange world as a result of the radical social transformation. It helped them rejuvenate, recharge, and then reenter a society that was fraught with pressure.

Clients' sexual beliefs, as well as their uncontrollable sexual desires, pursuit of sexual pleasure, release of pressure, belief in culturally defined sex roles, and passionate sex, play a crucial part in determining condom use in sexual encounters. A convergence of these issues has practical implications for HIV intervention methods. Since men believe that sex has to be passionate and spontaneous, the power of sexual impulse can override condom use or self-protection. As men in Mexico stated, the passion climbed so much that they were not interested in condoms or in protecting themselves (Carrillo 2002, 247).

In Dalian, the educational booklets the local CDC prepared and disseminated consisted of only one line of advice, urging safe sex and monogamous relationships. The rest of the pamphlet was about the perils and transmission routes of HIV. The underlying belief was that as long as people learned about HIV, their knowledge would help them transform their sexual behaviors. The current intervention message failed to address people's conceptualizations of sex, their use of intuition to assess sexual partners, and their emphasis on spontaneity, irrationality, and uncontrollable sexual passion. In this sense, the HIV intervention messages were divorced from local sexual culture. Unless the messages were adapted to their cultural environment and sensitive to the cultural context, HIV prevention work would not be successful or effective. This analysis sets the stage for the next chapter where I will focus on recommendations for future strategies for HIV prevention.

AFTERWORD

DURING MY FIELDWORK IN DALIAN, I CONDUCTED research in local non-governmental organizations (NGOs), the local Centers for Disease Control and Prevention (CDC), and the local Red Cross. My research has shown that they emphasize a message of safe sex and faithful monogamous marital relationships. The undergirding assumption is that if individuals are equipped with the knowledge that HIV is dangerous and lethal, they will modify and alter their sexual behaviors. The crux of the message is that it is the individuals' responsibility to protect themselves.

As my research shows, the individualistic approach emphasizing rational decision making based on knowledge does not work. It is insensitive to the larger sexual culture that shapes people's sexual behaviors in China. Researchers such as Parish and Pan also argued that an HIV intervention program that focuses on individual responsibilities is ineffective in China (Parish and Pan 2006). As they observed, "individual" in either ideology or practice did not exist in traditional Chinese culture (Parish and Pan 2006, 209). They argued that in this cultural background, where people equated individualism with selfishness, presenting HIV infection as an illness harmful to one's own life is ineffective (Parish and Pan 2006, 209–10). Furthermore, as shown in Chapter 6, clients' interpretation of sex as irrational and spontaneous is at odds with the HIV prevention message that is predicated on rational and self-controlled sexual behaviors (see also Carrillo 2002).

In this book, I have argued that it is the complexity of sexual culture, gender dynamics, and cultural beliefs that an effective intervention program has to consider. By delving into HIV prevention work and intervention designs, this chapter demonstrates that an effective HIV intervention must be sensitive to the broader contexts and frameworks, wherein sexual activities take place, and take into consideration the nuances of the social, cultural, and economic factors that inform individuals' behaviors and beliefs. Indeed, education messages that lack sensitivity to the local sexual culture are doomed to fail.

HIGH-RISK BEHAVIORS INSTEAD OF HIGH-RISK GROUPS

Chapter 2 has analyzed and critiqued the harmful effect of categorizations of high-risk groups instead of high-risk behaviors in HIV intervention work. Research on HIV prevention has proven that a high-risk-group approach has failed in controlling the epidemic (Elias 1991). Indeed, if it is only the high-risk groups who are identified and blamed for the spread of HIV, people who do not self-identify as a member of the high-risk groups do not perceive themselves at risk of infection (Piot and Laga 1991; Seeley et al. 1994; Seidel 1993). Hence, intervention messages cannot reach more people unless they focus on high-risk behaviors such as having multiple sexual partners and engaging in unsafe sex. Emphasis on high-risk behaviors can help people identify their risky actions and thereby perceive possibilities of infection (Messersmith et al. 2000). In view of Chapter 2, efforts to stop HIV transmission must include a critique of the hegemonic construction of risk groups.

HOSTESSES

As illustrated in the book, imbalances in gender power and financial resources prevent hostesses, and women in general, from negotiating safe sex (see also Holland et al. 1992b). Research elsewhere such as Manchester, South Africa, and London has shown that gender power dynamics have led many young women to experience sexual initiation through coercion and force (Campbell et al. 2001; Holland et al. 1991, 1992a, 1994a). Similar cases of physical violence and coerced condomless sex were also recorded in Chapters 4 and 5.

Because low condom use and high sexually transmitted disease (STD) infection rates pose potential HIV risks, researchers around the world have recommended peer education to equip female sex workers with negotiation skills. Indeed, it seems that negotiation skills and distribution of reliable and high-quality condoms are indispensable (see also Qu et al. 2002).[1] A few Chinese provinces hit hard by the AIDS epidemic, such as Yunnan and Guangxi, have programs to offer free condoms and health education to women working in the entertainment industry but on a limited basis (Liu 2001 et al.; Qu et al. 2002). However, it is critical that negotiation takes place from a power base, which is illustrated in Chapter 5.

In my interviews with the local Red Cross in 2006, I was told that it was extremely difficult to work with hostesses in China. One of the directors in the local Red Cross said,

It's impossible for us to openly conduct the HIV/AIDS intervention work. We dare not work with gay groups or sex workers because we're afraid of public and media criticisms. Because of the low HIV prevalence in Dalian, the city government continues to rely on anti-vice campaigns and imprisonment to stop the phenomenon of sex work and wipe out this disease. In Heilongjiang province, the Harbin Center for Disease Control (CDC) conducted a peer education class for sex workers. That event invited countrywide criticisms, as the activity was recognizing sex workers' legal identities. This event is our counter-example to avoid. We have to do things with low publicity [*di diao*].

As the director stated, public and media criticisms created an insurmountable barrier to work with gay groups or sex workers. The setbacks experienced by the Harbin CDC constituted a foil against which the protocols for the Dalian Red Cross were set. Since sex workers are illegal in China, intervention activities with them were seen to acknowledge their legal identities. Thus working with sex workers through a peer education class subjected the Harbin CDC to overwhelming criticisms.

As the director stated, because the city of Dalian has a low-HIV-prevalence rate, the city government continued to depend on harsh crackdowns, anti-vice campaigns, and imprisonment of sex workers.[2] As I related in my previous book (T. Zheng 2009), these methods were far from effective in curbing prostitution. Rather, they exacerbated the exploitation of, and violence against, female sex workers. These harsh policies not only had an adverse effect on sex workers' lives but also thwarted various agencies' intervention efforts.

The director told me that under pressure from their funding source—the Netherlands Red Cross—they conducted one training class for hostesses. The trainees were provided by the local NGO I had been working with. Because of lack of funding, the local NGO was not capable of financing such a training class. They had to rely on Red Cross funding to accomplish the task. I asked the director, "What was the content of the training?" He answered, "We taught them [hostesses] that contraction of AIDS is a death sentence that kills in ten years. We trained them to say the following to the clients if the clients refused to use condoms: 'AIDS is a disease that sentences a person to death. Even if you don't have the disease, you can't be sure that I don't have it. We cannot sacrifice our lives for momentary pleasure.' We taught them that sex has to be accompanied by condoms. It's very easy to impart this knowledge. It only took us ten minutes to finish the lecture."

His explanations shocked me. He did not understand that the questions they trained the hostesses to ask were far removed from the reality of these women's lives. Could they expect the hostesses to warn their clients who were the source of their livelihood that they might be the carriers of HIV? Would the hostesses be stupid enough to use this terrifying message to drive away their clients and leave themselves vulnerable to physical violence? What does this kind of training reveal about the trainers' attitudes toward hostesses and clients?

For the trainers to train the hostesses to caution their clients that they might carry HIV and possibly transmit the virus to the clients, treats the clients as victims and the hostesses as perpetrators. The purpose of the training, in this sense, lies in protecting clients' sexual health, rather than the hostesses'.

Training sexual negotiation has to be sensitive to women's situations. The following example illustrates an effective program developed elsewhere (Travers and Bennett 1996). The first task is to assess the most serious concerns and needs of the women at risk. The second undertaking is to understand women's loss of control and coping strategies. The final stage of skill training is specific, practical, and informative. The necessity for practice and rehearsal is stressed. For example, condom negotiation with potential partners is role-played. Furthermore, women need support in their continued use of new skills. Researchers emphasize that any programs that fail to acknowledge social restrictions to the implementation of new behavioral strategies are doomed to failure (Travers and Bennett 1996). In other words, programs require more than information and need to take into account complex factors that maintain certain behavioral patterns.

However, interventions based on negotiation skills may not be effective because, as illustrated in Chapter 5, hostesses may prioritize economic survival with condomless sex over health concerns. As shown in Chapter 5, hostesses have a continuum of risks to worry about, such as police arrests, physical violence, economic deprivation, and power asymmetry in their relationships with clients. Regular customers are their indispensable survival means, which they aspire to cling to rather than let go, even if it means risk of infections. Health, in this case, will be compromised and sacrificed for survival or a better living. HIV intervention work, to be useful and effective, needs to take these sociocultural factors into consideration. Furthermore, because hostesses' economic dependence on clients influences their condom use, effective intervention strategies should take

into account the economic aspect of the relationship and design programs
to help reduce poverty among hostesses.

Collective efficacy can be another effective social action (see Bandura
1997, 2005). Collective efficacy is defined as a group's belief in its ability
to take action and demand condom use. It has been documented among
Calabar sex workers in Nigeria, who cooperate to ensure consistent condom use with clients (Esu-Williams 1995; Oladosu et al. 2001). Sex
workers in Senegal also work together to remind one another, to help one
another to ensure safe sex, and to trick clients by using their legs instead
of vagina during condomless sex (Renaud 1997). Thailand's universal
condom policy helped prostitutes generate collective efficacy (Koetswang
and Ford 1999).

Researchers have demonstrated that initiatives for collective efficacy
rely heavily on preexisting networks of support and solidarity between
sex workers and other community members (Wojcicki 2001; Catherine
Campbell 2000). As they have observed, such approaches have promoted
a more open recognition of sex work as a profession, reduced the stigma
and discrimination sex workers face, and encouraged them to organize
themselves openly in groups dedicated to protecting their interests in an
assertive and public way. Collective activities are open and often noisy,
including composing and singing songs referring to the dangers of AIDS
and their determination to combat it (Catherine Campbell 2000).

The collective efficacy approach recognizes the agency of women who
make decisions and take actions to protect themselves. If local agencies,
as pointed out by the director of the local Red Cross, cannot work with
hostesses because of their illegal identity, then we have to resort to other
local NGOs to facilitate collective efficacy among hostesses, such as the
organization of Ziteng in Beijing.[3] However, no matter how effective
the local NGOs can be in fostering collective efficacy, an intervention
program developed only for women is still not effective on a large scale
because women lack sufficient power. Educational programs must also
be designed to change men's attitudes toward condoms and intervene in
men's sexual behaviors (see also Qu et al. 2002).

CLIENTS

HIV intervention work needs to consider client influences, power and
control, and such factors as economic dependence that adversely affect
women's ability to manage protective behavior during sex exchange. As
noted in this book, confirmed by other researchers, the most common

reason sex workers did not use condoms was customers' refusal (Qu et al. 2002).

Prevention work should focus on male sex consumers and terminate media portrayal of clients as victims of sex workers' infections. As demonstrated in Chapter 2, hostesses and other immoral women were portrayed as the carriers of the HIV virus, transmitting it to the "general population." This obscured the role of men, especially male sex consumers, who enjoyed the culturally prescribed male privilege of having multiple sexual partners and wielded their sexual power to abjure condom use. As seen in Chapters 4 and 5, these clients made the decision not to use condoms for various reasons, thereby transmitting diseases to their female sexual partners, in this case, hostesses.

This was also confirmed by research conducted on a global scale. In Australia, North America, Thailand, and Great Britain, research has shown that clients pose a threat of HIV infection to sex workers through unprotected sex (Day 1990; Knodel and Pramualratana 1994; Perkins et al. 1994; Plant 1993; Sacks 1996; Waddell 1996b). In China, Zheng et al.'s research among patients attending STD clinics showed that it was a large number of men who put themselves at risk for HIV by engaging in risky sexual behaviors (Zheng et al. 2000). For many women, however, their risk was the result of their partners' sexual behaviors rather than their own. Indeed, it was the hostesses who risked being infected by clients as a result of condomless sex. Hence, Zheng et al. (2000) appeal to the kind of prevention messages that target not only men and women but also the couple.

Intervention work should focus on the risks between regular clients and hostesses. Regular clients and hostesses go through a courtship process, whereby trust, romance, and affection are established over time. As a result, condom use is abjured. This finding is confirmed by research on sex workers around the world (Kerriga et al. 2003). It points to the urgent need to develop specialized education materials and intervention strategies to target hostesses and their regular clients.

Since a myriad of complicated cultural, political, and economic factors affect customers' refusal of condom use, as shown in Chapter 4, education programs should be devised to alter men's attitudes toward condoms and modify their risky behaviors. This work is much more creative and challenging than just to provide information. As demonstrated earlier, since clients believe that all STDs are curable, they are fearless in taking risks. In view of this situation, HIV intervention should carry the message that not all STDs can be cured.

Moreover, client-focused intervention programs have to heed that clients' actions and beliefs are influenced by their group membership and their peer culture. Research elsewhere has confirmed that group behaviors are influenced by peer norms, especially regarding sexual behaviors (Campbell et al. 2001; Reed and Weinberg 1984). For instance, studies of American college students have shown that discussions of safe sex with friends strongly influenced their practices of safe sex (Lear 1995).

Although these researchers have indicated that peer norms assist in their adoption of safe sex, my research has shown the opposite. This finding echoes the research conducted by Fisher and Misovich (1992), who have argued that in most cases, peer norms encourage risk. For instance, Thai men typically consume sexual services and large doses of alcohol as part of friendship parties (Maticka-Tyndale 1997). Young men in South Africa face peer pressure to engage in multiple sexual relationships without condoms, which places their sexual health at risk (Campbell et al. 2001; Holland et al. 1990; Wight 1994). Among clients in my research, peer norms function to promote risky sexual behavior, whereby clients strive to appear fearless and brave.

A peer education program, developed by Dube and Wilson (1996) in Zimbabwe was based on the social and psychological principle that people are more likely to change their perceptions and behaviors if they perceive their peers committed to the change (Lewin 1958). Peer involvement and peer pressure can trigger both positive and negative effects. On the negative side, peer pressure poses impediments to safety as peers encourage one another to participate in risky activities. On the positive side, peer involvement facilitates safety if peers endorse norms of risk reduction such as condom use. Hence, peer pressure and influence are important factors to heed in intervention programs. To offset the adverse effect of peer norms, it is worthwhile to implement peer education programs wherein respected peers function as facilitators to encourage safe sex (see Campbell et al. 2001; Serovich and Greene 1997; Tran et al. 2006).[4]

Research in other countries has demonstrated that using male peer groups to foster risk reduction in sexual contexts increases their consistent condom use (Fisher and Fisher 1992; J. A. Kelly 1992; Maticka-Tyndale 1991, 1992). Clients' political views, gender perceptions, sexual identities, and peer culture influence men to jettison condoms. Changing peer culture has to start with changing peer behaviors and worldviews. Peer education can be conducted in male consumers' workplaces by selected members of the workforce who have received the training to facilitate discussion, in an effort to eventually modify peer behaviors and peer culture.

FUNDING AND LOCAL NGOS

The issue of lack of funding for local NGOs should be amended. Financial strains have made it difficult for local NGOs to carry out the intervention work. During the time when I was working with one of the local NGOs, they wished to run a peer education class for hostesses but could not put it into practice because of insufficient funds. Why did local NGOs lack funding? As mentioned in the introductory chapter, it was not only because the central government mistrusted and exerted control over them but also because, in this case, members in the group were gay. The head of the local Red Cross said, "Local NGOs were comprised of gay members. They dare not mention anything gay in their NGO titles. In fact, they chose titles such as 'Consultation Agency of AIDS' to cover up their gay identity. They conducted work in the name of AIDS prevention working groups. Otherwise the government and people would not understand and media would criticize it so much that they would have to stop their work. Since their real identities are covered up, the Red Cross and the local CDC can work with them. Otherwise we would suffer from rampant criticisms."

During the time when I was working with the local NGO, we were trying to initiate a peer education class for hostesses in cooperation with the local Red Cross. I proposed to the director of the Red Cross that I would be responsible for bringing along the women, and they would provide compensation fees for them. My proposal was turned down. The director said, "The relationships between our agencies and local NGOs are very complicated. Because the NGO you represent happens to be the enemy of a couple of other NGOs that we have relationships with, we cannot cooperate with you. If we did, the other NGOs would be unhappy."

However intricate the relationship was, I felt that the HIV intervention work should take precedence. If we let politics get in the way, how can intervention work be accomplished? The NGO members told me that local government agencies and GONGOs, or government-organized nongovernmental organizations, did not care about the real intervention work. Rather, they only wished to complete the quota set up by the source of their funding and profit from the grant. One local NGO member said,

> The responsibility of the local CDC is to hang up flyers on the walls of karaoke bars and distribute free condoms. However, in reality, their major

responsibility is to earn money through offering people immune vaccines and HIV tests. They only dispense the flyers once a year during World AIDS Day when they have to perform something. They pretty much just hang them up on the walls and check off one item on the list. It's very superficially done. They don't care whether people really get the message or not. As an NGO, we are unable to apply directly for global funds. We have to apply to the local CDC and receive it from them. They allotted us a very small amount for our intervention activities, with extremely strict conditions. Afterwards, credit for our intervention work belonged solely to the CDC. The CDC does not recognize local NGOs. They treat us as their volunteers. So all our work goes to their credit. Because we don't have the legal rights to apply or receive global funds, all local NGOs have to look for chances to cooperate with the CDC.

Over the years, the local NGO members I worked with had developed resentments toward these local government agencies and GONGOs such as the Red Cross. One of the local NGO members adamantly criticized the corruption of the Red Cross and the local CDC in front of me: "The local Red Cross only spent ten cents [*fen*] on each printed little book, but they put down 1 dollar [*yuan*] in their reports to the Netherlands Red Cross. They embezzled a great deal of money by printing out thousands of these little books. Look at their employees and leaders—they are driving expensive cars and holding cell phones that are worth thousands of dollars. Look at their remodeled buildings and equipped infrastructure such as huge-screen TV and DVD players. All those came from the funding that was supposed to be spent on AIDS prevention work."

Local NGO members complained to me that the local CDC and Red Cross had abused a huge amount of the grant for their own use. When I was acting as the consultant for a local NGO, I followed their members to the local CDC and Red Cross to talk to the officials. The NGO members directed my attention to their elaborately furnished offices, brand-new buildings, private cars, and their expensive cell phones.

During the time when I worked with the local NGO as a consultant, I learned about their financial dependence on the Red Cross and the local CDC. Because only the local government agencies and GONGOs provided funding, local NGOs, to apply for any kind of finding, had to cooperate with either the Red Cross or the local CDC. Since there were a number of local NGOs in the city and limited funds, they competed with each other to curry favor with the local Red Cross and the local CDC so that they could be chosen for cooperation.

It is axiomatic that the political environment has severely hindered intervention. The eligibility requirement for applications for funding has led to competition, and at times, vicious fighting, rather than cooperation, among local NGOs. However, local NGOs' legal identities were not acknowledged but treated as dependent volunteers.

Moreover, the gay identity of local NGO members constitutes another barrier to their intervention work, as the local government agencies and GONGOs would only "adopt" them as volunteers if the names of their organizations suggested no gay identity. This means that these agencies would abandon them anytime for fear of countrywide criticisms should controversy over gay issues arise, leaving them penniless and completely stunted. It also means that any of these NGOs' political efforts or campaigns related to gay issues could be severely curtailed and terminated.

It is evident that this political environment should be altered and modified. The government should be more progressive and open about local NGOs, rather than being suspicious of them. This means to loosen the eligibility criteria for funding and provide local NGOs with a space to survive so that joint efforts instead of intense competition between groups could lead to more successful and effective intervention.

MASS MEDIA

Marketing of condoms with a focus on STD prevention has proved extremely successful around the world, in Zaire, Zambia, Zimbabwe, Nigeria, the Dominican Republic, and Indonesia (Adetunji et al. 2003; Agha 1997; Fajans, Ford, and Wirawan 1995; Messersmith et al. 2000).[5] Condom brands such as Dua Lima, Right Time, and Gold Circle that were promoted in the marketing campaign have been the most widely cited by respondents (Fajans, Ford, and Wirawan 1995, 24; Messersmith et al. 2000). In Africa, the Zambian Social Marketing program has advertised and promoted Maximum condoms since 1992, using the slogan "Strong for maximum protection, sensitive for maximum pleasure" (Agha 1997). Research has shown that respondents who have heard the brand-advertising message are significantly more likely to use condoms during sexual activities. In Nigeria, sex workers who have been exposed to two or more sources of advertising for Gold Circle and Cool condom brands are about two times more likely to use condoms consistently than those who have not seen any advertisements (Oladosu et al. 2001). In Uganda, large-scale condom marketing began in the mid-1990s and has increased condom use dramatically (Hearst and Chen 2004). Ugandans now use more condoms than before (Hearst and Chen 2004).[6] In Cameroon (Klein 2001), the "100 percent Jeunes" project has promoted consistent condom use with regular partners in particular, through a mass media campaign, radio call-in shows, newspaper, and radio dramas.

In Asia, Thailand's experience has also shown that with an intense 100 percent condom use promotion program, condom use in brothels has risen from about 14 percent to more than 90 percent (Rojanapithayakorn and Hanenberg 1996), and 89 percent of indirect sex workers have used condoms with paying clients, as compared with 19 percent with nonpaying clients (Mills et al. 1997). Cambodia, the country that originally had the highest HIV rate in Asia, with a high proportion of transmission occurring through commercial sex (Chanpong et al. 2001) also instituted a 100 percent condom program, and STD rates and HIV prevalence has fallen substantially among sex workers (Hearst and Chen 2004).

It has been proven worldwide that advertising of condoms helps increase condom brand awareness, strengthen clients' willingness to use condoms regularly, and increase consistency of condom use, as it improves attitudes toward condoms and awareness of condom efficacy. Indeed, results have strongly suggested that condom use has increased substantially as a result of condom marketing, promotion, and distribution activities (Agha 1997). The finding of a strong association between the brand-advertising message and condom use shows that condom advertising is an effective way to increase condom use. On a global level, mass media have also proved to be most effective in encouraging couples to discuss sensitive issues such as condom use.

While mass media campaigns do not have as strong an effect on a particular individual to use condoms as do peer educators or providers, they have a substantial effect at the population level because of their considerable reach. Therefore, condom promotion and distribution campaigns should not only focus on bars, hotels, and other entertainment places but also should target the community and the society as a whole (see also Zheng et al. 2000).[7] Various means such as posters, brochures, and outreach educational workers at entertainment places and hotels, have proved effective and useful in other parts of the world (Gossett and Warshaw 1992). Companies, government offices, and other workplaces are also indispensable sites to target since those are clients' working places.

Mass media are pervasive enough to penetrate all nooks and crannies of society, including the clients. In general, clients are a difficult group to target, as they do not expose themselves or congregate in a specific place as hostesses do. They are like a revolving door, coming and going swiftly, in and out of entertainment places and hotels. The ephemeral and amorphous nature of this group makes it difficult to access them. Yet mass media can reach the population of clients more effectively.

Condom campaigns should be sensitive to the peer culture of male sex consumers and pay special heed to the most effective message to encourage men to carry condoms. As illustrated in the book, clients believe in spontaneous and passionate sex. They hate the moment ruined by stopping to look for condoms. This means that the timing and place of sex are not planned but on the spur of the moment or impulsive. Therefore, availability of condoms is critical for actual condom use. Successful campaigns have to promote carrying condoms at all times and address men's sexual beliefs. Marketing interventions should make condoms attractive; reduce the shyness, embarrassment, and the stigma of condom use;, and help men overcome barriers to carrying condoms. Vending machines should be made available in public places, and free condoms should also be provided at social settings and entertainment sites.[8] Shaping HIV prevention messages according to local expectations about sex will help individuals enact their sexual desires while minimizing risks for HIV infection.

Male clients in my research are different from Latino men who are reported more likely to use condoms for contraception. Research of Latino men has shown that they are less likely to use condoms if another form of birth control is used, but more inclined to use condoms as a method of contraception (Yeakley and Gant 1997). This is also true in other countries such as Norway (Mahler 1996), Cameroon (Calves 1999), Madagascar (Silva et al. 2003), and Nigeria (Rossem et al. 2000).[9] Hence, researchers from these areas appealed for intervention messages stressing condoms for contraception rather than vehicles for disease prevention. However, clients in my research represented a contrary case. They did not use or rarely used condoms for contraceptive purposes before and after marriage. Emergency pills, intrauterine devises, abortions, and contraceptive pills were used to prevent pregnancies.

In addition, as shown in Chapters 3, 4, and 6, condoms in China have been associated with such authoritarian, top-down, mandatory, and coercive family planning programs that it has been perceived as intrusive in people's everyday lives and hence oppressive, which is one reason clients rejected condom use—as a political act of defiance. In view of this situation, different from other countries, prevention messages in China should eschew, rather than focus on, the definition of condoms as a means of contraception. Indeed, condom use should be disentangled from associations with government authority, health authorities, population control, and unfaithfulness (see also Kaler 2004).[10] Instead, condoms should be associated with staying healthy and caring for the sexual partner.

Condom promotion should also break the association between condoms and unfaithfulness. Research elsewhere has shown widespread resistance to condom use for its association with a lack of trust (Blecher 1995; Carole Campbell 1995; Cohen and Trussel 1996; Maharaj 2004; Worth 1989). As explicated in the book, since unfaithfulness and a lack of trust are associated with condom use, condom use suggests infidelity and disease. As hostesses do not want to jeopardize their relationships with client lovers, it is difficult for them to suggest condom use. Possible suspicions could create tension, anger, confrontation, and, ultimately, loss of financial sources for the hostesses.

Condom promotion should not target prostitutes specifically, as it would only reinforce the association between condom use and prostitution. Regular clients and hostesses form a romantic and intimate-lover relationship, in which case, they would not perceive themselves at risk and hence not choose condom use. Therefore, condom promotion should construct and establish a positive image of wise and masculine male condom users that men would aspire to emulate.

Mass media efforts should not only counter the negative feelings people associate with condom use but also should emphasize safe sex with regular partners, including married partners (see also Knodel and Pramualratana 1994; Messersmith et al. 2000; Qu et al. 2002). Resonating with the book, research in other countries has also shown that sexual activity and condom use are determined by a complex assessment of each relationship and vary greatly, depending on the nature of the relationship, the meaning and level of trust in that relationship, and the relative power and ability to assess and control the sexual behavior of the partner (Messersmith et al. 2000). Condom use decreases as a relationship grows more stable and intimate over time (Macaluso et al. 2000). As the relationship status between hostesses and clients changes from new to regular, condoms are used less. Therefore, mass media communications should address the complexities inherent in different kinds of sexual relationships while promoting safe sexual behavior. Such a campaign is more likely to have a greater effect on HIV and STD prevention than information programs targeting specific groups.

Condom promotion should also destigmatize females who purchase or carry condoms. Young hostesses who purchase condoms are identified by shop clerks as prostitutes. Indeed, for many years in China, possession of condoms had been the evidence used to arrest a woman for engaging in prostitution (X. Ren 1999). Similarly, in South Africa, women often do not have condoms available and make few efforts to gain knowledge of their partners' sexual histories, as it would be tantamount to admitting to themselves and society that they plan to engage in sex. As in China, social pressures encourage women not to engage in sex, but those that do are expected to do so in the confines of serious and trusting relationships. This emphasis on serious relationships encourages premature trust of partners and therefore nonuse of condoms (Hillier et al. 1998; Holland et al. 1990; Ingham et al. 1991). Since this social environment stigmatizes females procuring and carrying condoms, condom promotion campaigns should debunk the negative image attached to young, unmarried women purchasing and carrying condoms. A positive image of female condom users or carriers should be established.

Female condoms are a conspicuous lacuna in China, as they are not present in pharmacies, adult health shops, or supermarkets. The unavailability and high cost of female condoms have made it difficult for hostesses to use them. Mass media campaigns promoting female condoms have demonstrated their extraordinary effect in altering people's conservative opinions about female condoms (see Agha et al. 2001). For instance, a study in Sichuan has discovered that a significant number of sex workers were unwilling to accept female condoms. However, following a pilot training program funded by China–UK alliance, there has been a dramatic increase of female condom use (Yan Wang 2007). In Tanzania, mass media campaigns implemented during 1999 through radio and newspapers have had a significantly positive effect on intentions of men and women to use female condoms (Agha et al. 2001). In Zambia, since the mass marketing, studies have indicated that awareness of female condoms was as high as 80 percent in 1999 (Meekers 1999). Given the lack of discussion of condoms between sexual partners, mass media can remove the barrier and facilitate discussions of safe sex (Agha et al. 2001). Mass media promotion of female condoms motivated sexual partners to discuss female condoms, and the discussion exerted a strong influence on their intentions to use them.

Last, mass media campaigns should be consistently carried out over the years. According to the local NGO members in my research, the World AIDS Day on December 1 was the only time when the local government agencies and health authorities held health education activities or awareness-raising campaigns, such as exhibitions about HIV/AIDS in certain places and broadcasts of educational programs on HIV/AIDS through TV and radio (see also Shen et al. 2004). To make mass media campaigns effective, it is not enough to concentrate only on a couple of days per year. Ongoing and consistent mass media campaigns all throughout the year are essential and crucial to increasing society's understanding of HIV/AIDS.

COMMUNITY INVOLVEMENT

Researchers such as James Pfeiffer (2004) have emphasized the danger of not involving community participation in the AIDS prevention efforts. As he observes, in a central Mozambican community, the social marketing of condoms is implemented at the expense of community dialogue and reduces community participation in confronting the AIDS epidemic. Indeed, as he points out, the Pentecostal and Independent churches in the community have expanded rapidly across the region and now represent most of the population. Despite the advertising campaign developed to sell condoms, the churches have clashed with this campaign and have spread a contrasting message about sexuality and risky behavior. The lesson in this Mozambican community is that community involvement must be elicited and tapped into to ensure the success of the campaign.

Involving entertainment owners or managers in supporting the campaign constitutes one aspect of community involvement. Examples in Thailand and Cambodia have demonstrated that establishment-based condom use policies can have a dramatic effect. Support by the brothel owner, pimp, manager, or others who control the life and work of sex workers is essential to prevent economic reprisal if a woman refuses a client who is not willing to wear a condom (Wawer et al. 1996). Setting standards and regulating compliance are important at entertainment places.

Thailand has enjoyed great success in implementing 100 percent condom use in brothels with support from the owners and local governments (Fajans, Ford, and Wirawan 1995). The Thai government mandates that brothel owners enforce condom use in every paid sex act (Koetswang and Ford 1999). Uncooperative owners are identified through STD surveillance among sex workers and clients and receive sanctions. Since health officials worked with brothel owners on instituting 100 percent condom use in all brothels in 1992, condom use has reached more than 90 percent in commercial sexual encounters (Hanenberg et al. 1994), and the proportion of men visiting sex workers has fallen by half (Mills et al. 1997; Phoolcharoen 1998). The government did not directly discourage commercial sex, but mandatory condom use and the awareness of risk caused many men to give up paying for sex. Thai men also reduced the number of their unpaid casual partners (Mills et al. 1997). Rates of STDs fell rapidly, and HIV incidence and prevalence are declining.

The successful case of Thailand reveals that the involvement of owners in condom promotion is important, as they often have great influence over the practices of clients and hostesses. Although there is such a policy in China, it has not been effectively implemented. The local government and health authorities should monitor and ensure that sex establishment owners make condoms available to sex workers and clients. This strategy will help create an environment wherein intervention work is shared by community members rather than belonging to certain individuals.

Community involvement cannot ignore the importance of clinics. Research on STDs in China shows that efforts should include patient education during regular clinic visits (Zheng et al. 2000). My interviews with clinic doctors have suggested that they accept male infidelity as the norm. Instead of actively emphasizing the importance of informing sexual partners, they just rely on patients to tell their partners about their infections. As one can predict, many of them leave the clinic with drugs and conceal their infections from their partners. The same issue is also reported in studies of clinics in Brazil (Giffin and Lowndes 1999). To remedy this situation, clinic doctors should join HIV intervention efforts to combat partner infidelity and stress the importance of informing sexual partners of infections.

Community involvement should also include cooperation with local agencies. Because the government mistrusts AIDS NGOs, NGOs have been under strict surveillance by the government, and NGO leaders have been subjected to police harassment and police arrest. Leaders such as Wan Yanhai and Hu Jia were subjected to multiple detainments and imprisonment, not to mention everyday police surveillance and harassment. As police harassment escalated, in June 2008, Wan Yanhai sued local police for violating his freedom and human rights. He was also busy nullifying and condemning various rumors and accusations about his "conspiracies" against the central government on the AIDS working group e-mail list. Preoccupations with these daily nuisances run the risk of not only compromising the efficacy of local NGOS in HIV intervention efforts but also reining in leaders' endeavors in the cause.

In this aspect, Uganda is a successful example to emulate. Uganda had among the world's highest AIDS prevalence rates in the 1980s. The government responded to this situation with a determined approach involving all sectors of the society. More than seven hundred agencies in Uganda are cooperating with the government working on AIDS prevention, ranging from churches to NGOs to the military (Green et al. 2002; Hearst and Chen 2004). Since then HIV incidence has fallen substantially.

China should learn from these successful examples and integrate them with its own cultural, political, and economic specificities. The government should respond to AIDS decisively and cooperate with local NGOs. Local NGOs should be recognized and eligible for government funding to perform HIV intervention work effectively. National programs should have leadership from the highest levels and aim to be multisectored and multifaceted. Social marketing campaigns and national programs should seek to achieve broad public support, combat stigmatization, and emphasize condom use. Peer education programs should be sensitive to the special needs and peer culture of clients and hostesses to be effective. National programs should also seek out community involvement and support, including owners of entertainment places and clinics. These culturally conscious responses to the AIDS epidemic will have a promising and positive effect on curbing the HIV epidemic in China.

Notes

1. Three provinces in southern China have been identified as harboring the highest documented rates of HIV prevalence among sex workers—Guangxi (10 percent), Guangdong (3 percent), and Yunnan (5 percent; Zhang et al. 2002, 51). In Chongqing, the HIV infection rate due to sexual transmission rose to 23 percent in 2002 from 10 percent a few years earlier, and in Yunnan, the rate was 21 percent in 2004. Data for 2004 also reported 27 percent unknown routes of transmission, of which many may be sexual (Zhang et al. 2002, 51).

2. As the most prominent HIV/AIDS activist working in China, Wan Yanhai has worked considerably with the LGBTQ (lesbian, gay, bisexual, transgender, and queer) population in both the context of HIV/AIDS prevention and human rights. In 2002, the government kidnapped Dr. Wan from a gay and lesbian film festival in Beijing for his involvement in exposing a major blood-selling scandal in Henan province, where HIV infections skyrocketed after local government officials and businessmen used unsanitary methods to obtain and sell blood from peasants. Despite this and other instances of government detainment, Wan Yanhai has continued to fight for the rights of all those affected by HIV/AIDS in China.

3. It was reported that data collection surveillance system is only based on primary risk groups such as detainees in detention centers (Zhang et al. 2002, 56).

4. By the end of 2004, nationwide surveillance system has increased to 247 sites in thirty-one provinces. The surveillance approach involves serological testing among the five "high-risk groups" twice a year—IDUs and female sex workers in detention centers, public STD clinic patients, truck drivers, and antenatal clinic attendees (Zhang and Anxun 2002, 54).

5. As Dechamp and Couzin (2006, 148) state, patients must pay for tests, drugs, and other treatment, as local government has to use testing and treatment as income-generating medical services.

6. As Saich (2006, 39) points out, local leaders have not followed the central government leader to pay visits to AIDS patients or publicly embrace AIDS patients.

7. I interviewed condom company owners about their marketing strategies of condoms, and I obtained from the family planning office the sex education CD for newly married couples.

8. Interview questions also focused on the use of alcohol in their sexual interactions; their perception of the beginnings of a sexual relationship; their views

about who should initiate a sexual relationship and how they communicated with their partners about their likes and dislikes during sex; what told them that an interaction was sexual; their understanding of safe and unsafe sex; their views about whether there were differences between sex behaviors with a wife and with a prostitute or a lover; whether they carried condoms; whether their partners had condoms available; whether they assessed the dangers of disease before engaging in sex; assessment of education and social class as a basis for trust; which party normally spoke about protection; where they obtained condoms when they needed them; what prevented them from asking for a condom in an encounter; what had happened to them when they had asked for protection; how emotional involvement related to use of protection; and so on.

CHAPTER 1

1. Mann (1997), Ko (1994), and Bray (1997) have recorded significant changes over time in the position of women. Their works have traced how women were situated in the three-tiered hierarchy (the imperial state, elite family, and peasant household) and how they advanced or suffered in such a patriarchal system. They have employed the late imperial state-elite-peasant triad to examine women's domesticity, life course, reproduction, motherhood, work, feelings, education, and property. Each of women's roles was considered in relations to orthodox literati culture of Confucianism and prominent popular culture of Buddhism and Daosim.

2. This concept originates from Lao Zi's *Dao De Jing*. Tao is the cosmos, the truth, the way, and the life. Lao Zi wrote, "Tao of Heaven resembles the stretching of a bow. The mighty it humbles, the lowly it exalts. They who have abundance it diminishes and gives to them who have need. That is Dao of heaven; it depletes those who abound, and completes those who lack."

3. Exceptions of this general theme are reflected in the homosexual culture and prostitutes in Tang and Song poetry.

4. This theory was abolished since Chinese medicine associated some disease with copulation.

5. Almost all sex manuals taught men the therapeutical intercourse postures; choice of sex partners; adjustment of sexual behaviors according to seasonal changes, dates, and time; and recipes of yang-strengthening food. This was to prepare for the conception of healthy children with their wives through strong ejaculation.

6. It was argued that the semen was latent for years and would be reactivated during each intercourse. The remnants would enter the vagina through the bloodstream.

7. As instructed, one should first thoroughly wash a fresh cecum, soak it in water containing a thousandth part of mercuric chloride, and then degrease it in alcohol. Mechanical dilation and a rubber ring to seal off one end would make the condom ready for use. The cervical cap, only invented at the turn of the century in Europe, was also recommended in the 1920s. The circulated literature

also detailed other birth-control methods such as vaginal sponges, pessaries, interuterine devices, acidic powders, and jellies (Dikotter 1995).

8. The health of the patriarchal state included an ongoing concern with eugenics.

9. Deng Yingchao was Premier Zhou Enlai's wife.

10. M. Nie (2005, 93) contends that the official line on permitting termination at any stage of pregnancy except in sex-selective abortion does not necessarily accord with historical Chinese values and practices. He argues that forced termination of a pregnancy sets up an example in which the state justifies any means to achieve good causes. This, as he observes, has alienated Chinese people from the government and aggravated the conflicts between people and the state (M. Nie 2005, 220). In terms of people's standpoints on abortion, he argues that Chinese people exhibit great diversity, ambivalence, uncertainty, and inconsistency regarding the moral status of the fetus and the moral nature of abortion (M. Nie 2005, 132).

11. See J. Wei (2002). While there is no statistical data on the percentage of women who use liquid condoms, in my ethnographic research on karaoke bar hostesses during the summer of 2005, of the twelve women I interviewed, all favored the use of liquid condoms.

12. J. Wang (2006) points out that the Chinese court system believes that rape within marriage does not constitute a crime as long as it is conducted within a legal marriage. For instance, in one case, the wife, Yao, requested a divorce from her husband Bai Junfeng (J. Pan 2005). After the divorce request, the two were physically separated. Yao lived in her parents' house, waiting for the divorce agreement to be reached. One day, her husband came to her parents' house, pressed her down, and used physical violence to force intercourse. Yao continuously resisted during the process, but to no avail. Yao lost consciousness due to his physical violence and was sent to the hospital. She did not come back to life until the doctor intervened with medical assistance. Although the husband employed physical violence against his wife and the wife vehemently resisted throughout the process, the verdict of the people's court was that the husband's behavior did not constitute rape because it was carried out within matrimony (J. Pan 2005).

CHAPTER 2

1. "That it is a punishment for deviant behavior and that it threatens the innocent—these two notions about AIDS are hardly in contradiction" (Sontag 1989, 64).

2. Along with Emily Martin, I recognize that there are multiple discourses in a cultural field. Jing Jun, for instance, has approached the construction of HIV/AIDS in the Chinese media in a different angle from mine. His focuses on the incitement of fear and terror of AIDS patients in the media, and the lack of social trust inherent in the processes of such representation (Jing 2006).

3. Jing Jun (2006) argues that the Chinese news media criminalizes people with AIDS. He points out that the rumors of "AIDS criminals" reflect a crisis of

social trust in current China. On the basis of an extremely thin layer of evidence, the rumors are amplified, sensationalized, and distorted to serve the purpose of criminalizing marginalized sufferers of AIDS. The failure of government officials to criticize the role of media indicates that government officials are not fully aware that China's crisis of social trust is the driving force of the panics associated with the AIDS rumors.

4. In the end, it had to take place in the dining hall of the Bureau of Hygiene, and only thirteen of the twenty-seven invited cadres showed up (Hu 2005).

5. Also screened are prostitutes, pregnant women, visitors to STD clinic, drug addicts, blood donors, and long-distance truck drivers. By 2002, the country has established 158 national sentinel surveillance sites. One can assume that mandatory testing and surveillance are China's attempt at disease prevention and control.

6. Buckley (1997) criticizes Japan's response to the AIDS epidemic for blaming the outside for the infection, stigmatizing homosexuality, and valorizing the nuclear heterosexual household within Japan. Buckley states that the fact that infected blood in Japan came primarily from the United States contributes to the constructed foreignness of the disease. As she contends, the Japanese official figures always separate foreigners from the Japanese, resulting in the stigmatization of the illegal immigrant women in the entertainment and sex industry. Japan also emphasizes avoiding gay sex and sex work and presents homosexuals, foreigners, and drug users as threats to the happy, risk-free heterosexual family. The media contend that it is the body of the mother and the heart of the family that act as a zone of sanctity and safety.

7. This law, titled *zuigaorenminfayuan guanyu dangqian banli liumang anjianzhong jutiyingyong falu de ruoganwenti de jieda* (The Supreme Court's Answer to Questions in Laws on Hooligan Cases), was published on November 2, 1984.

8. Premier Wen Jiabao proposed nine AIDS intervention measures, one of which was cracking down on criminal activities, including prostitution (Hao 2005).

9. The Japanese media and public health programs construct the migrant sex workers as the source of HIV/AIDS (Buckley 1997). As a result, they are policed by blood testing and periodic raids to "protect" their Japanese male clients from them.

10. It was argued that the semen was latent for years and would be reactivated during each intercourse. The remnants would enter the vagina through the bloodstream.

11. Dikotter first expressed this idea (Dikotter 1995).

12. "Talk of condoms and clean needles is felt to be tantamount to condoning and abetting illicit sex, illegal chemicals" (Sontag 1989, 75).

13. Sex education in China echoes that in Vietnam, South Korea, and other countries, where the young are protected from sexual knowledge and early sexual intercourse (Blanc 2004; Cheng 2005).

14. Sontag states that epidemic disease usually elicits a call to ban the entry of foreigners and immigrants. "Xenophobic propaganda has always depicted immigrants as bearers of disease" (Sontag 1989, 62).

CHAPTER 3

1. Trainings on condom marketing to "high-risk" groups took place in a limited number of cities such as Xinjiang, Sichuan, and Yunnan, funded by the China/UK project (Lin et al. 2002). The World Health Organization also started the pilot project of "100 percent condom use" in Wuhan city and Jingjiang County in 2000. Vice Premier and Minister of Health Madam Wu Yi, remarked specifically on the promising results of the 100 percent CUP strategy: "The 100 percent CUP in Hubei and Hunan have proven to be measures that have significantly decreased the incidence of sexually transmitted infections and played an active role in prevention and control of HIV/AIDS, and are worthy of further expansion to other provinces" (Lijuan Wang 2007).

2. (Hunter 2005, 91).

3. Seventy-six percent of these students expressed confidence in the "one stable partner" strategy.

4. In Nigeria (Messersmith et al. 2000), for instance, the increasing acceptability of condoms is indicated by frequent references to condoms in the popular press and by vigorous social marketing campaigns for such brands as "Right Time" and "Gold Circle." Zambia's committee has sponsored newspaper advertisements and comic books featuring a fanged yellow blob that says, "'I am the AIDS virus. I am very small. I am very dangerous." More than five hundred Zambian high schools now have an extracurricular activity called the anti-AIDS club, whose members sponsor lectures, visit AIDS patients, and publish poetry about viruses and monogamy. In Zambia, for instance, the reported incidence of sexually transmitted diseases has declined 15 percent each in 1998, 1999, and 2000. Condom sales are increasing throughout Africa, especially in countries with social-marketing programs (Tierney 1990).

5. Despite the assumption that marketing condom use and safe sex would encourage sexual activities, a plethora of studies conducted worldwide have concluded otherwise. For instance, on the basis of years of research, the 1999 policy of the American Medical Association concluded that promotion of safer sex is effective in delaying sex in teenagers, and abstinence-only programs have limited value (Stine 2007, 257). In another three-year and ten-month study in Switzerland, a public education campaign promoting condom use was found to be effective in increasing condom use and did not raise the ratio of adolescents who were sexually active. The research revealed that condom use among seventeen- to thirty-year-old people increased from 8 percent to 52 percent. By contrast, the ratio of adolescents from sixteen to nineteen years old who had sexual intercourse did not increase over that same period. Three other studies have drawn the same conclusion that promoting condom use does not boost sexual activities. Deborah Sellers et al.'s (1994) study and Sally Guttermacher et al.'s (1997) research concluded that the promotion and distribution of condoms did not increase sexual activity among adolescents. In another study that measured the number of condoms students took and the subsequent changes in sexual behavior, Kirby's research of ten Seattle high schools also showed that making condoms available

through vending machines and school clinics did not lead to an increase in sexual activity (Kirby et al. 1998; Stine 2007, 257).

6. Couples, after marriage, receive a CD that contains knowledge on sexual activities, including sex postures, contraception, and how to initiate and conduct sex.

7. According to China's State Quality Inspection Bureau, 30 percent of China-produced condoms failed to meet the industry standards—fifteen of the fifty condom brands surveyed failed strength tests (Qu et al. 2002). Researchers in China reported a condom slippage rate of 20 percent and a breakage rate of 13 percent among female sex workers, much higher than the 2 percent and 1.9 percent, respectively, in the United States, and 0.8 percent to 1.9 percent and 0.8 percent to 4.7 percent in Mexico, the Philippines, and the Dominican Republic (Qu et al. 2002, 274).

8. Some university students point out that sexual behavior exists on campus. They argue that it is normal for college students to engage in sex and they should not suppress their desires as long as the desires are released with certain restraints. To them, it is indispensable that condoms enter campus because it protects them from disease and pregnancy. Condom distribution will not only educate these students about risky behaviors, but also dissipate other students' curiosity about contraceptives and direct them to purchase the appropriate ones. These students consider it social progress and a sign of civilization to usher condoms into school because it promotes safe sex (Zhang 2006).

9. The advocates believed that condom advertisements should not be prohibited as long as images and words are supervised and obscene contents are avoided. The youth should be educated in both sex morality and condom use because it is more pressing for youth to learn how to protect themselves from diseases. Lifting the ban will usher in new ideas and promote social progress. Although they recognized the conflict between moral education and practical education, they insist that condom use does not loosen moral education. Rather, both educations should be emphasized. Although students receive education in morality at the school, youth outside of school are deprived of any sort of education, especially because condom advertisements are prohibited (Yanchun Li 2004).

10. This regulation was issued by six departments of the government, including the Hygiene Department, Family Planning Committee, Food and Medicine Supervision Bureau, Industrial and Commercial Bureau, Broadcast Bureau, and Quality Supervision Bureau.

CHAPTER 4

1. Courtesan houses or public places where courtesans were summoned as professional entertainers formed an integral part of the official and business routine where social relations of officials, literati, artists, and merchants were conducted. Every official entertained his close colleagues—superiors and inferiors and merchants—to conclude or to negotiate deals. An official could ensure his promotion by introducing his superior or an influential politician to a discreetly chosen

courtesan, and by the same means, a merchant could obtain a much-needed credit or an important order.

2. Mayfair Yang (1999b) states that according to the men, before liberation, men could gain economic and political power, but in the Maoist society, they were stuck in socialist work units earning the same meager wages as women. The party-state constantly watched over them, stifling their personal ambitions and prohibiting them from speaking their own minds. That led to men's feminization and lack of initiative and creativity. Susan Brownell (2000) argues that the state support of women and the state patriarch in the Maoist period is felt as emasculation by many Chinese men. Zheng Yefu published a book in 1995 asserting that the Maoist state's alliance with liberated women stifled men's ability to discover their own strength and led to their feminization (Zheng cited in M. Yang 1999a). He celebrated the new sexual discrimination against women in China by the market as a "rectification of past distortions in state planning" (Zheng cited in M. Yang 1999a, 54). If men feel impotent with powerful women but satisfied with suffering women, then as Brownell (2000) attests, this may explain the prevalent image of the suffering women who sacrifice for the nation.

3. Before the New Culture movement, women sacrificed for the patriarchs of the family. During the Maoist era, once again women were called on to sacrifice for the patriarchal state. Whereas a suffering woman could redeem national shame and promote an image of masculine success during the 1980s, they also created impotent and feminized men. The discourse on the suffering and self-sacrificing Chinese woman was really about men: "Women suffer because men are impotent to right the injustices done to men through their women" (Brownell 2000). Susan Brownell observed that the Qing dynasty celebrated female sacrifice as a symbol of morality and the strength of the empire; in other words, female bodily suffering was the key to the welfare of the empire. During the Qing dynasty, the male complement to the suffering female was the filial son, the embodiment of masculinity and potency. After the fall of the Qing, the common masculine image in China became not the filial but the unfilial and ultimately impotent son. "This theme of female suffering and male impotence continued in Chinese nationalism until the 1990s" (Brownell 2000). During the New Culture movement, reformers had attributed men's impotence to the family patriarch. After the New Culture movement, during the Maoist era, the family patriarch was destroyed and replaced by the state patriarch. During this era, the state's suppression of female sexuality was a means to control men's sexuality. Men later blamed their impotence on the state's advocacy of female liberation to control males. This changed during the 1990s; men could imagine themselves to be cosmopolitan and global by emulating the economically successful Taiwanese and Japanese businessmen in their consumption of women. Thus, at various times during the twentieth century, the cause of men's impotence was colonialism, political manipulation, and new and challenging economic systems.

4. Mayfair Yang observes that Chinese entrepreneurs, instead of taking offense against the Taiwanese and Japanese businessmen who had taken Chinese mistresses, simply emulated them and took mistresses themselves (Brownell 2000).

While young entrepreneurial men recovered their economic and sexual potency, older retired cadres were faced with impotence (Brownell 2000). So devastated was this group, there was an upsurge in the market for tonics to reinvigorate their sexual life. Here, the link between politics, economics, and sexuality is drawn. Men with economic and political power become sexually potent, whereas men who have lost such power feel emasculated by the market reforms. In Judith Farquhar's analysis of the 1994 film *Ermo*, she notes that Ermo's husband, a retired Communist cadre, formerly the village head and still called "Chief," is impotent (Farquhar 2002). In the new rural entrepreneurism, everyone must do business to live, and the chief's bureaucratic skills are useless. Ermo's neighbor, Blindman, on the contrary, owns a truck and has access to urban distribution networks through which Ermo sells her noodles. Ermo and Blindman, who are linked by business interests, have an affair. Ermo is married to a useless past, but the future that Blindman offers is corrupted by his eagerness to make her his dependent. "Outside of Ermo's troubled home she sees only lonely competition and relations based on the exchange of cash" (Farquhar 2002). Farquhar argues that in China's "rapid plunge into global modernity" Ermo is trapped between "an impotent old husband and a profligate new lover, neither of whom offers her a viable alternative to the unending labor of domestic production and petty commerce." This should be read as "a denunciation of the failures of the past and the banality of the future for a China that has committed itself at every level, even the most domestic, to millennial capitalist relations of production. Impotence in this domain is serious indeed." Thus, Susan Brownell (2000, 230) attests that "masculinity is related to state power, nationalist ideology, the free market, and the marriage/sex markets. The current situation has unleashed an entrepreneurial masculinity that is apparently proceeding hand in hand with the return of male privilege and female disadvantage."

5. My ethnography of the karaoke bars (T. Zheng 2009) not only displays the kind of entrepreneurial masculinity sought there but also demonstrates the responding femininity performed there. I argue that the hostesses take advantage of the clients' use of them and perform an obedient and promiscuous role to satisfy the clients. In return, hostesses redistribute the clients' social and economic resources and claim for themselves a cosmopolitan image.

6. See Messersmith et al. (2000). Men rarely use condoms with wives and casual partners. As a result, more than half of the male samples reported having acquired an STD from casual partners who were not sex workers.

7. These seven factories were located in Guangzhou, Guilin, Shanghai, Qingdao, Tianjin, Dalian, Shenyang.

8. Kaler (2004) states that in many countries in Africa, family planning programs bear a long history of suspicions that they are intended to stop population growth in brutal ways. For instance, in Nigeria, rumors circulate that family planning is part of a neoimperialist plan by the Western world to oppress Africans. In Tanzania, common perception holds that family planning is part of a postcolonial white conspiracy to keep African populations small. In Tanzania, the government family planning campaign failed because of rumors that family

planning was intended by the mainland Tanzanian government to make women barren. Among Cambodian refugees in Thai refugee camps, rumors spread that the provision of family planning methods was intended to finish what the Khmer Rouge started—to destroy the population, this time by offering Depo-Provera to women. Women on aboriginal reserves in northern Canada harbor suspicions that the government doctors who urge them to adopt oral and injected contraceptives have hidden agendas involving wiping out demographically fragile native nations.

9. For instance, Knodel and Pramualratana (1994) contends in his research that the major obstacle to common use of condoms is that most Thai men dislike using condoms, whether in the context of commercial or noncommercial sex relations and whether for prophylactic or contraceptive purposes. He argues that in general, Thai men believe condoms detract from the pleasure of sexual intercourse. That men dislike condoms surfaced in almost all the male focus group discussions and in most of the interviews with in-depth male respondents. He notes that, if condoms are used, it is because there is a compelling reason to do so, despite that it detracts from sexual pleasure. Furthermore, no one indicated that he or other men liked using condoms, and no one explicitly denied that condoms reduced, at least somewhat, sexual pleasure. He points out that Thai men believe condom use reduces the enjoyment of sex has been documented in numerous studies using both qualitative and quantitative methodologies. For instance, among samples selected in 1992, of three different groups targeted for a study dealing with issues related to HIV/AIDS in Thailand, 63 percent of male truck drivers, 74 percent of low-income urban adult men, and 72 percent of low-income urban male adolescents agreed that condoms reduce the pleasure of sex. Many men agreed that "condoms reduce sexual sensitivity and pleasure," and that "during sexual intercourse condoms are an interference."

10. Scholars have emphasized the prevalence of "sexual silence" in Mexican culture and Latin culture (Carrillo 2002; Gomez and Van Oss 1996). Sexual silence refers to the phenomenon in which individuals are acculturated to avoid talking openly about sex and maintain indirect and veiled communication about sex. As Carrillo notes, it is a "widespread method used to keep transgressive sexual behavior under wraps in order to maintain the appearance of normality. They [people] indeed discovered how one can transgress and simultaneously comply with cultural expectations, how nonnormative behaviors can be carried out without triggering negative social consequences. This included, as well, learning that certain forms of sexual joking were allowed in order to refer to sexual matters in good company or to safely express interest or desire to potential sexual partners" (Carrillo 2002; Gomez and Van Oss 1996).

11. Research has shown that men in southern Africa regularly do not want to wear condoms and that men have their masculinity intimately tied with flesh-to-flesh sex (Webb 1997).

12. Calves (1999), in her study in Cameroon, states that while the image of a boy purchasing condoms is mixed and can be positive at times, girls who buy or are seen carrying condoms are classified as prostitutes. She concludes that this finding

explains why female adolescents are significantly less likely to buy condoms than their male counterparts. Not only are females less likely to buy condoms, but condom procurement through female friends is also stigmatized. While it is common for male adolescents to obtain condoms from friends, female adolescents, especially younger ones, are reluctant to do so because they do not want to disclose that they are sexually active for fear of ruining their reputation. Overall, the negative image attached to female adolescents procuring condoms represents a serious barrier to condom purchase and procurement among girls. Calves thus proclaims that HIV/AIDS prevention programs and reproductive health promotion campaigns should focus on destigmatizing condom use among females.

CHAPTER 5

1. For ethnographic studies on hostesses, see T. Zheng (2009).
2. The Maoist state initiated the household registration system in 1958, which outlawed rural migration through the management of resource distribution and thereby established a two-tier urban-rural caste system. Concomitant with the Communist broad-based restructuring of society, the "peasantry," as a derogative cultural category and revolutionary mainstay was further refined and concretized.
3. I made friends with these two hostesses while I was living with hostesses and working with them as a hostess during my ethnographic fieldwork from 1999 to 2002. For the detailed ethnography about the hostesses, please see (T. Zheng 2009).
4. Please refer to my first book on karaoke bar hostesses for a detailed account of the violent, hostile, and antagonistic working environment of hostesses (T. Zheng 2009). Although hostesses fall into a gray area—the law does not clearly identify them as either illegal or legal—in everyday practice "hostesses" provide illegal erotic services and hence are the major target of the antiprostitution campaign. Police raids make them both legally and socially vulnerable. If their sexual services are disclosed by the clients to the police, they would be subject to extreme humiliation, arrest, handsome fines, and incarceration. In their everyday lives, the local police constitute their daily fear and terror. Because the police wield arbitrary power, the hostesses are obliged to obey their sexual demands without monetary compensation. Local officials not only sexually and economically exploit hostesses but also keep "spy hostesses" as their personal harem. The anti-vice campaign also allows the bar owners to impose more severe regulations and disciplines on the hostesses who otherwise would operate in a more laissez-faire manner. Because the state's anti-vice policy is manipulated and usurped by local officials and bar owners for their own ends, leading to a violent working environment for the hostesses, hostesses do not disclose their real identity, which makes it more convenient for men to be violent toward them and even to murder them.
5. Wojcicki and Malala (2001) in their research on female sex workers in Johannesburg, Wojcicki and Malala (2001) argue that these women choose not to

get tested (and if they are tested, they do not return for the results) as a coping strategy to handle the stressors in their lives: poverty, violence, and discrimination. They argue that these coping strategies are not healthy but do not indicate that women are in denial concerning the epidemic. Rather, they contend that as researchers, we should try to understand why women choose not to get tested and not return for their results from the women's perspectives.

6. In a previous article, I (T. Zheng 2008b) dealt with the complexity of hostesses' collective actions. On the one hand, large-scale mobilization is inhibited by their unstable coalitions (or complete disaffiliation), the incentive to select noncooperative strategies, and the costs and risks associated with collective action. These factors make it impossible to overcome the barriers of personal interest and danger that impede large-scale collective mobilization. On the other hand, hostesses form unstable small-scale group cliques and rely on these networks for help and security. These informal networks are facilitated and enabled by blood relationships, common rural background, native-place, and mutual benefits. However, these informal networks are transient and temporary due to the internal competition and the costs and risks of protests.

7. Kerriga et al.'s study (2003) shows that sex workers had an average of two client dates per week, but only three sexual partners in the past two months, suggesting that the majority of these sexual partners were regular paying partners. She points out that this finding confirms the urgent need to develop specialized educational materials and intervention strategies to address the potential increased risk of HIV and STD transmission between female sex workers and their regular paying partners.

8. Kapumba et al. (1991) points out in her research that focus-group participants believed in the "love potion" effects of drying agents in dry sex, particularly their ability to attract and keep a sexual partner. Tonga women in Zambia applied the drying agents in their vaginas. The Zambian women commonly used "love medicines" made from plant products to initiate and maintain a relationship, as well as to compete with potential rivals. Kapumba illustrates that study participants reported that they were reluctant to use condoms because condoms were considered a physical barrier and would block the love potion effects of the agents. They believed that condom use would interfere with the effectiveness of drying agents. Therefore, Zimbabwean health professionals said that any attempt to alter women's use of traditional agents must be sensitive to these cultural factors.

9. Studies have shown that the location of the event might be correlated with some important client or relationship attribute influencing condom use.

10. Hansen, Lopez-Iftikhar, and Alegaria (2002) point out that kissing was associated with closeness to clients.

11. As Civic and Wilson (1996) contends, women reported physical problems stemming from the effect that agents had on men's libido. Yet women said, "The man keeps coming back. It will be hurting you, but he will be enjoying it." "The man wants sex all night which will cause you to have a swollen vagina." Group participants in all groups answered that condoms frequently break when used in

conjunction with drying agents. The primary deterrent to using condoms, cited by two groups, was that condoms block some of the "love potion" effects of the agents. "If you use condoms, the magic doesn't work right—you need skin to skin contact." Several participants noted that drying agents are used more often with steady partners, where condoms tended not to be used, so that the agents could "work their magic."

12. Oladosu et al. (2001) contend that condom negotiation during sex exchange encounters typically involve economic and survival considerations. As they point out, prior research has shown that nonuse of condoms can be a critical bargaining tool used by women to procure additional drugs or money from exchange partners (Hansen, Lopez-Iftikhar, and Alegaria 2002; Wojcicki and Malala 2001). They comment that they still know very little about the content and context of these critical discussions. They argue that for HIV/STI prevention programs targeting female sex exchangers to be successful, a greater understanding of women's control or lack of control over sexual practices in various social contexts is required.

13. Wojcicki and Malala (2001) conducted interviews with fifty female sex workers in the Johannesburg area and argued that sex workers capitalized on clients' reluctance to use condoms in sexual exchanges. They also examined other elements of the sex industry that contributed to unsafe sex, such as competition between women for clients and violence in the industry. They suggested that we should recognize the agency of sex workers and thus lessen the stigma and discrimination that sex workers face at the hands of clients, pimps/managers, police, and health care workers. They also commented that public health programs should incorporate a component that emphasizes reducing the stigma and discrimination that sex workers face.

14. Research in Greece (Paxson 2002, 319) has shown that for a woman to introduce condoms into marital relations could be seen as shedding doubt on her own or her spouse's sexual fidelity.

15. Researchers Wojcicki and Malala (2001) and Campbell (2000) have contended the agency of female sex workers. Wojcicki and Malala (2001) argue that by emphasizing the victimhood of sex workers and women in general, past studies have failed to recognize that women are decision makers and actors. They state that past studies contribute to an overall negative discourse that continues to stigmatize sex workers. This focus on "powerlessness" obscures everyday decisions and actions that women are engaged in. Catherine Campbell (2000) notes that sex workers have various means of empowering themselves in dealing with difficult environments in Carletonville, South Africa. Campbell emphasizes the micro decision making that sex workers are making in light of structural inequities.

16. Sanders (2004) argues that sex workers rationalize the outcomes relating to non-condom use by relying on the excellent health care services available, their knowledge of what to do should such an occasion arise, and peer support. The legalities around selling sex place women's mental and physical well-being at risk. Pressure to hide their work, to live a double life, to fabricate stories to their families and

partners to avoid stigma, and to marginalize result in significant psychological stress. By legitimating sex work as a profession, the structural inequalities that leave many women vulnerable could be addressed, enabling women to organize themselves in public and in private without fear of committing an offense. Protective relationships with the police could also be established and resources could move away from criminalization, fines, arrests, court appearances, probation, and imprisonment through antisocial behavior orders.

CHAPTER 6

1. Parker writes, "Ethnographically grounded descriptive and analytical research on the social and cultural construction of sexual meanings provides important insights to the representations shaping HIV-related risk and offers the basis for the development of culturally sensitive and culturally appropriate, community-based HIV/AIDS prevention programs" (Parker 2001, 168).

2. My client interviewees are forty to fifty years old. Their high schools and colleges during the conservative 1970s and 1980s did not provide sex education. I also interviewed young adults in their early twenties as a comparison. A twenty-one-year-old young adult answered this way, "In high school, when teachers were supposed to teach sex education, they asked all the boys out of the classroom and only taught girls about menstruation. Then they asked all the girls out, then taught the boys about sexual organs. In college, sex education was an optional class, only scheduled during weekends. Many students took different classes to fulfill this requirement. Some took the class, but professors never called rolls. So students rarely attended class, but just recited some stuff to pass the exam to fulfill the credits."

3. Gordon writes, "For women, therefore, heterosexual relations are always intense, frightening, high-risk situation is which ought, if a woman has any sense of self-preservation, to be carefully calculated. These calculations call for weapons of resistance, which may include sexual denial . . . and pregnancy itself" (Gordon 1979, 127).

4. Furth (1994) contends that the promotion of longevity was in accord with Daoist goals of self-cultivation. She points to the struggle between the positive tradition of healthy sexual hygiene practiced during the early empire (Han-Tang) and the puritanical ideology of neo-Confucianists who sought to suppress teachings about sex. Medieval Chinese correlated sex with serious goals of life and death, linking health, spirituality, and social purposes. The erotic could be seen as a vehicle for individual self-transformation and sagehood. Dominion over one's own body, the bodies of women, and the body politic came together to realize "true men." Sexual intercourse was used as a method to attain longevity as lords and sages. Control of body and self to transcend the body's ordinary limitations and achieve immortality was the goal. Self-discipline and self-denial were cultivated as virtues. The ideal man was a person not easily moved by women. The ideal man was a person whose self-control was manifest "in a dispassionate

inner calm and an outwardly youthful appearance even in old age" (Furth 1994, 145).

5. The strategic importance of balancing this form of "animalism" with "social convention" is caught up in the deployment of social etiquette aimed at discouraging abnormal sexual conduct (Sigley 2001).

6. They state that, although both young men and young women are constantly warned about "the dangers of unrestrained sexual activity," it is young women who are at the focus of the attention.

7. Paxson writes, "Far from disrupting this paradox, family-planning rhetoric has tended to reinforce women's accountability, demonstrating that 'ideologies of modernization, so often thought to challenge traditional gender roles and relations and, in particular, to benefit women, have just as often reinforced the 'traditional' sexual division of labor" (Paxson 2002, 322).

8. Sexual joking, in the Mexican society, is a socially acceptable way of establishing veiled communication about sexual topics—communication that does not break the rules of sexual silence (Carrillo 2002, 150–51).

9. His access to these services at sauna bars hinged on his wealth. That is, the high consumption rate of these services determined that only the wealthy could afford to be regular patrons.

AFTERWORD

1. Research in Bali, Indonesia, showed that about half of the clients would agree to use a condom on requests (Fajans, Ford, and Wirawan 1995, 120).

2. For the city's political policy of crackdown, anti-vice campaign, and imprisonment of sex workers, please see my previous work for detailed description and analysis (Zheng 2009).

3. Zi Teng is a nongovernmental organization focusing on sex workers' concerns. Participants in the organization range from social workers, labor activists, researchers specializing in women studies, to church workers. These workers care and concern about the interest and basic rights of women. In their mission statement, Zi Teng states, "We believe that all women, regardless of their profession, social classes, religion, or races, have the same basic human rights, that they are equal and entitled to fair and equal treatment in the legal and judicial system, that nobody should be oppressed against, that all people should live dignity."

4. Researchers have even argued that if possible, peer educators can change peer norms of defining masculinity as accepting multiple sexual partners.

5. In Nigeria (Messersmith et al. 2000), for instance, the increasing acceptability of condoms is indicated by frequent references to condoms in the popular press and by vigorous social marketing campaigns for such brands of condoms as Right Time, Gold Circle, and others.

6. Ugandans now use more condoms than before, and there has been a substantial drop in the numbers of casual sex partners (Hearst and Chen 2004).

7. Zheng et al. (2000) have promoted that mass media campaigns target everyone in the society.

8. According to Goodkind (1997), one of the reasons for the recent rise in condom use is the increased availability of condoms in Vietnam.

9. In Cameroon (Calves 1999), condom use for pregnancy prevention is more acceptable among youth than condom use for STDs.

10. In Malawi (Kaler 2004), people have negative feelings toward condoms because condoms are associated with population control. Rumor goes that people who do not succumb to AIDS are at risk from the very thing they use to avoid AIDS, condoms.

REFERENCES

Abdool, Karim S. S., Karim Abdool Q, Sankar N Preston-Whyte. 1992. Reasons for lack of condom use among high school students. *South African Medical Journal* 82:107–10.

Adams, Vincanne, and Stacy Pigg. 2005. Sex in development: Science, sexuality, and morality in global perspective. Durham, NC: Duke University Press.

Adetunji, Jacob, and Dominique Meekers. 2003. Social marketing and communications for health consistency in condom use in the context of HIV/AIDS in Zimbabwe. PSI Research Division Working Paper No. 19. PSI (Population Services International). Washington, DC.

Agha, Sohail. 1997. Sexual activity and condom use in Lusaka, Zambia. PSI Research Division Working Paper No. 6. No. 19. PSI (Population Services International). Washington, DC.

Agha, Sohail, and Ronan Van Rossem. 2001. The impact of mass media campaigns on intentions to use the female condom in Tanzania. PSI Research Division Working Paper No. 44. No. 19. PSI (Population Services International). Washington, DC.

Ai, Pin. 2004. Aizi xiaotou panxing: Rendaodajizhihou de zhidukaoliang (Sentenced AIDS thieves: Measure of system after moral attack). *Fazhi yu xinwen* (*Law and News*) 4:34–36.

Amadora-Nolasco, Fiscalina, Renae Alburo, Elmira Judy Aguilar, and Wenda Trevathan. 2001. Knowledge, perception of risk for HIV, and condom use. *AIDS and Behavior* 5 (4): 319–30.

An, Ni. 2006. Anquantao yehui taozou nuxing jiankang (Condoms can strip a woman of health). *Qianlong Xinwen Wang*, April 10.

Anagnost, Ann. 1995. A surfeit of bodies: Population and the rationality of the state in post-Mao China. In *Conceiving the New World Order*, ed. Faye Ginsburg and Rayna Repp. Berkeley: University of California Press.

Andors, Phillis. 1983. *The unfinished liberation of Chinese women, 1949–1980*. Bloomington: Indiana University Press.

Anonymous. 1964. *Nongcun weisheng shouce* (*Hygiene handbook in the countryside*). Nanchang: Jiangxi Renmin Chubanshe.

———. 1991. Biyunhuan wenti (Questions on IUD). *Women of China* 10:42.

———. 2002. *Anquantao tuxian anquan wenti* (*The issue of safety of condoms*). http://www.jynk.com, August 32.

———. 2004a. 2004 wenjuan (Survey in 2004). *Dazhong wenzhai* (*Popular Digest*) 11 (47): 16–17.

———. 2004b. Xingqu shoucang (Sex interests). *Shenghuo yu jiankang* (*Life and Health*) 2:34.

———. 2005a. Publication Statistics (Faxing Tongji). *Bosom Friend* (*Zhiyin*) 3:12.

————. 2005b. Zhuanjia yuce: Zhongguo anquantao shichang qianli chao baiyiyuan (Experts predicted that the market potential of condoms in China exceeded tens of billions of yuan). http://info.china.alibaba.com, December 21.

————. 2006a. Jiankang shenghuo yidiantong (Help on health and life). *Dushi Zhufu (Hers)* 7:116.

————. 2006b. Lian aizibing renshu shangsheng 46.67% (HIV infection increasing rate is 46.67% in Dalian). *Dongbei Xinwen,* November 29.

————. 2006c. Zhongguo Aizibing Ershinian (1985–2005) (Twenty Years of AIDS in China). Sohu.com.

Aral, Sevgi. O., and Janet S. St. Lawrence. 2002. The ecology of sex work and drug use in Saratov Oblast, Russia. *Sexually Transmitted Diseases* 29:789–805.

Baer, Hans A., Merrill Singer, and Ida Susser. 1997. *Medical anthropology and the world system.* Westport, CT: Bergin Garvey.

Bandura, Albert. 1977. *Social learning theory.* Englewood Cliffs, NJ: Prentice-Hall.

Bandura, Albert. 1997. *Self-efficacy: The exercise of control.* New York: W. H. Freeman.

————. 2005. Guide for creating self-efficacy scales. In *Self-efficacy beliefs of adolescents,* ed. Frank Pajares and Tim Urdan. Greenwich, CT: Information Age.

Bankole, Akinrinola. 1999. Book reviews. *Studies in Family Planning* 30:89–92.

Bankole, Akinrinola, G. Rodriguez, and C.F. Westoff. 1996. Mass media messages and reproductive behavior in Nigeria. *Journal of Biosocial Science* 28:227–39.

Barlow, Tani. 1994. Theorizing woman: Funu, Guojia, Jiating. In *Body, subject and power in China,* ed. Angela Zito and Tani E. Barlow. Chicago: University of Chicago Press.

Barme, Geremie. 1995. To screw foreigners is patriotic: China's avant-garde nationalist. *China Journal* 34:209–34.

Bastos, Cristiana. 1999. *Global response to AIDS.* Bloomington: University Indiana Press.

Basuki, Endang, Ivan Wolffers, Walter Deville, Noni Erlaini, Dorang Luhpuri, Rachmat Hargono, and Nuning Maskuri. 2002. Reasons for not using condoms among female sex workers in Indonesia. *AIDS Education Preview* 14:102–16.

Bin, Lang. 2003a. AIDS shiji youling (Century ghost). *Jiankang Guwen (Health Consultation)* 1–3 (124): 4–13.

————. 2003b. Butong renqun de yufang (Prevention for different groups). *Jiankang Guwen (Health Consultation)* 1–3 (124): 9.

————. 2003c. Shenmo shi Aizibing (What is AIDS). *Jiankang Guwen (Health Consultation)* 1–3 (124): 6–7.

Birkinshaw, Marie. 1989. *Social marketing for health.* Geneva: World Health Organization.

Blanc, Marie-Eve. 2004. Sex education for Vietnamese adolescents in the context of the HIV/AIDS epidemic: The NGOs, the school, the family and the civil society. In *Sexual cultures in East Asia,* ed. Evelyne Micollier, 241–62. New York: RoutledgeCurzon.

Blecher, Mark, M. Steinberg, W. Pick, M. Hennick, and N. Durcan. 1995. AIDS knowledge, attitudes, and practices among STD clinic attenders in the Cape Peninsula. *South African Medical Journal* 85 (12): 1261–86.

Bloor, Michael. 1995. *The sociology of HIV transmission*. London: Sage.

Bolton, Ralph, and Merrill Singer, eds. 1992. *Rethinking AIDS prevention: Cultural approaches*. Philadelphia: Gordon Breach Science.

Bond, George C., John Kreniske, Ida Susser, and Joan Vincent. 1997. *AIDS in Africa and the Caribbean*. Boulder, CO: Westview.

Brandt, Allan. 1988. AIDS: From social history to social policy. In *AIDS: The burdens of history*, ed. Daniel Fox and Elizabeth Fee, 141–71. Berkeley: University of California Press.

Bray, Francesca. 1997. *Technology and gender: Fabrics of power in Late Imperial China*. Berkeley: University of California Press.

Brownell, Susan. 1995. *Training the body for China: Sports in the moral order of the People's Republic*. Chicago: University of Chicago Press.

———. 2000. Gender and nationalism in China at the turn of the millennium. In *China briefing 2000*, ed. Tyrene White. Armonk, NY: M. E. Sharpe.

Buckley, Sandra. 1997. The foreign devil returns: Packaging sexual practice and risk in contemporary Japan. In *Sites of desire: Economics of pleasure*, ed. Lenore and Jolly Manderson, Margaret, 262–91. Chicago: University of Chicago Press.

Cai, Fang. 2000. *Zhongguo renkou wenti baogao (A Report of the Problem of the Chinese Population)*. Beijing: Shehui Kexue Wenxian Chubanshe.

Calves, Anne E. 1999. Condom use and risk perceptions among male and female adolescents in Cameroon. PSI Research Division Working Paper No. 22. PSI (Population Services International). Washington, DC.

Campbell, Carole. 1995. Male gender roles and sexuality: Implications for women's AIDS risk and prevention. *Social Science & Medicine* 41 (2): 197–210.

Campbell, Catherine. 2000. Selling sex in the time of AIDS: The psycho-social context of condom use by sex workers on a Southern African mine. *Social Science & Medicine* 50:479–94.

Campbell, Catherine, Zodwa Mzaidume, and B. Williams. 1998. Gender as an Obstacle to Condom Use: HIV Prevention amongst commerical sex-workers in a mining community. *Agenda* 39:50–59.

———, and Catherine MacPhail. 2001. I think condoms are good but, aai, I hate those things: Condom use among adolescents and young people in a Southern African Township. *Social Science & Medicine* 52:1613–27.

Cao, Xiaoyong. 2004. Jujue tongfang shi jiating baoli ma (Is Refusal to Have Sex Domestic Violence)? *Zhongguo Funu (Women of China)* 3 (621): 29.

Carrier, Joseph M. 1989. Sexual behavior and the spread of AIDS in Mexico. *Medical Anthropology* 10:129–42.

Carrier, Joseph M., and Rachel Magana. 1991. Use of ethnosexual data on men of Mexican origin for HIV/AIDS prevention programs. *Journal of Sexual Research* 28 (2): 189–202.

Carrillo, Hector. 2002. *The night is young*. Chicago: University of Chicago Press.

Chanpong, Gail Fraser, Maidy Putri, Sophal Oum, Ung Sam An, Mam Bunheng, Jeffrey Ashley, James R. Campbell, and Andrew L. Corwin. 2001. Prevalence of HIV infection in Cambodia: Implications for the future. *International Journal of STD and AIDS* 12 (6): 413–16.

Chatterjee, Partha. 1993. *The nation and its fragments*. Princeton, NJ: Princeton University Press.

Che, Yan, and John Cleland. 2003. Contraceptive use before and after marriage in Shanghai. *Studies in Family Planning* 34 (1): 44–52.

Chen, Jiali. 2002. Yangshi anquantao guanggao beipo linshi genggai (CCTV condom advertisements forced to be changed at the last minute). *Zhongxinwang* (*Chinese News Internet*), December 5.

Chen, Shi. 1958. Shengyu you jihua, shengchan jintou da (Planned birth, more energy for production). *Women of China* 5:30–31.

Chen, Yang. 2005. Shenmo shi anquan xingshenghuo (What is safe sex). *Jiankang Zhoubao* (*Health Weekly*) 1 (18): C3.

Cheng, Sealing. 2005. Popularising purity: Gender, sexuality, and nationalism in HIV/AIDS prevention for South Korean Youths. *Asia Pacific Viewpoint* 46 (1): 7–20.

Chiang, Mai. 2004. Brief introduction of school HIV/AIDS prevention education in China. International seminar/workshop on learning and empowering key issues in strategies for HIV/AIDS prevention, Thailand, March 1–5.

Ch'iu Lyle, Katherine. 1980. Report from China: Planned birth in Tianjin. *China Quarterly* 83:551–67.

Chu, Zhaorui, and Suide Shao. 2005. *Aizibing fangzhigongzuo keburonghuan* (*AIDS prevention work is pressing*). Zhongguo Aizibing Fangzhi (Prevention of HIV/AIDS in China). Beijing: Capital University of Medical Science.

Civic, Diane, and David Wilson. 1996. Dry sex in Zimbabwe and implications for condom use. *Social Science & Medicine* 42 (1): 91–95.

Clarke, Kamari Maxine. 1999. To reclaim Yoruba tradition is to reclaim our queens of Mother Africa: Recasting gender through mediated practices of the everyday. In *Feminist fields: ethnographic insights*, ed. Sally Cooper Cole, Rae Bridgman, and Heather Howard-Bobiwash. New York: Broadview Press.

Clatts, Michael. 1989. Ethnography and AIDS intervention in New York City: Life history as an ethnographic strategy. In *Community-based AIDS prevention, studies of intravenous drug users and their sexual partners*, ed. Michael Clatts. Rockville, MD: National Institute on Drug Abuse.

———. 1994. "All the king's horses and all the king's men": Some personal reflections on ten years of AIDS ethnography. *Human Organization* 53:93–95.

Cohen, Barney, and James Trussell. 1996. *Preventing and mitigating AIDS in sub-Saharan Africa: Research and data priorities for arresting AIDS in sub-Saharan Africa*. Washington, DC: National Academy Press.

Coleman, Patrick. 1988. Enter-educate: New word from Johns Hopkins. *JOICFP Review* 15:28–31.

Cook, James. 1996. Penetration and neocolonialism: The Shen Chong rape case and the anti-American Student Movement of 1946–47. *Republican China* 22 (1): 65–97.

Cusick, Linda. 1998. Female prostitution in Glasgow: Drug use and occupational sector. *Addiction Research* 6:115–30.

Day, Sophie. 1990. Prostitute women and the ideology of work in London. In *Culture and AIDS*, ed. Douglas A. Feldman, 93–110. New York: Praeger.

Day, Sophie, Helen Ward, and John Richard Harris. 1988. Prostitute women and public health. *British Medical Journal* 297:1585.

de Zalduondo Barbara, and Jean Maxius Bernard. 1995. Meanings and consequences of sexual-economic exchange. In *Conceiving sexuality: Approaches to sex research in a postmodern world*, ed. Richard G. Parker and John H. Gagnon, 155–80. New York: Routledge.

Dechamp, Jean-Francois, and Odilon Couzin. 2006. Access to HIV/AIDS treatment in China: Intellectual property rights, generics, and barriers to effective treatment. In *AIDS and social policy in China*, ed. Arthur Kleinman, Joan Kaufman, and Tony Saich, 125–51. Cambridge, MA: Harvard University Asia Center.

Dikotter, Frank. 1995. *Sex, culture, and modernity in China: Medical science and the construction of sexual identities in the Early Republican Period*. London: Hurst.

———. 2004. A history of sexually transmitted diseases in China. In *AIDS in Asia: The challenge ahead*, ed. Jal P. Narciin, 67–84. New Delhi: World Health Organization Regional Office for South-East Asia, Sage.

Dong, Bian. 1955. Zenyang renshi biyun wenti (How should we think about the problem of contraception). *Xin zhongguo funu* (*Women of New China*) 4 (66): 27.

———. 1965. Yingai quanmian de lijie jihua shengyu (We must understand family planning from all aspects). *Women of China* 12:30.

———. 1966a. Yinggaixiang laoren he nantongzhi xuanchuan jihuashengyu (We should broadcast family planning to the old and the men). *Women of China* 4:32.

———. 1966b. Zhege tou daidehao (It is great to take the lead). *Women of China* 1:25.

Dong, Jingmin. 2005. Shoudu jichang jianqi Aizibing Jiancedian (Surveillance was established at the capital airport). *Jiankang wenzhai bao* (*Health and Digest Newspaper*) April 3:8.

Dong, Tong. 1999. Xingbaojianpin dadande xianqi gaitoulai (Lifting the veil of sex health products). *Beijing Chenbao* (*Beijing Morning Newspaper*), November 30.

Dong, Xiaoci, and Ling Wang. 2003. Zhangfu Fubai laopo guan? "Furen geming" yinfa sikao (Husbands are corrupt, should the wives be responsible? "Wife revolution" induces thinking). *Zhongguo Funu* (*Women of China*), August (1): 14.

Douglas, Mary. 1991. Witchcraft and Leprosy: Two Strategies for Rejection. *Man* 26 (4): 723–36.

Du, Hailan. 2004. Aizibing fangzhi de zhengcefalu yudai gaishan (Room for improvement of AIDS policies and laws). *Fazhi ribao* (*Law Daily*), December 1:5.

Dube, N., and D. Wilson. 1996. Peer education programs among HIV-vulnerable communities in Southern Africa. In *HIV/AIDS Management in Southern Africa:*

Priorities for the Mining Industry, ed. Brian Williams and Catherine Campbell, 107–10. Johannesburg: Epidemiology Research Unit.

Elias, Christopher. 1991. Sexually transmitted diseases and the reproductive health of women in developing countries. Working Paper No. 5. Population Council, New York.

Elifson, Kirk W., Jacqueline Boles, William Darrow, and Claire Sterk. 1999. HIV Seroprevalence and risk factors among clients of female and male prostitutes. *Journal of Acquired Immune Deficiency Syndromes and Human Retrovirology* 20:195–200.

Epstein, Helen. 2003. AIDS in South Africa: The invisible cure. *The New York Review of Books* 50 (11).

Esu-Williams, Eka. 1995. Clients and commercial sex work. In *HIV and AIDS, the global interconnections*, ed. Elizabeth Reid, 91–99. West Hartford, CT: Kumarian Press.

Evans, Harriet. 1997. *Women and sexuality in China: Dominant discourse on female sexuality and gender since 1949*. London: Polity Press.

Evans, Harriet. 2002. Past, perfect or imperfect: Changing images of the ideal wife. In *Chinese Femininities, Chinese Masculinities*, ed. Jeffrey Wasserstrom, Susan Brownell, 335–60. Berkeley: University of California Press.

Fajans, Peter, Kathleen Ford, and Dewa Nyoman Wirawan. 1995. AIDS knowledge and risk behaviors among domestic clients of female sex workers in Bali, Indonesia. *Social Science & Medicine* 41:409–17.

Fan, Guiyu. 2001. *Zhonghua shengyu wenhua daolun* (*An Introduction to Chinese Birth Culture*). Beijing: Zhongguo renkou chubanshe (China Population Publishing House).

Fan, Hui. 2002. Anquantao guanggao mingnian zhengshi fangkai? (Will condom ads be opened up next year?). *Zhongguo xinwen wang* (*China News Net*), December 6.

Fang, Gang. 1996 Aizibing bijin zhongguo (AIDS impending China). *Women of China* (*Zhongguo Funu*) 1:24–27.

Farmer, Paul. 1992. *AIDS and accusation: Haiti and the geography of blame*. Berkeley: University of California Press.

———. 1999. *Infections and inequalities: The modern plagues*. Berkeley: University of California Press.

———. 2006. A biosocial understanding of China. In *AIDS and social policy in China*, ed. Arthur Kleinman, Joan Kaufman, and Tony Saich, x–xxii. Cambridge, MA: Harvard University Asia Center.

Farmer, Paul, Margaret Connors, and Janie Simmons, ed. 1996. *Women, poverty, and AIDS: Sex, Drugs, and structural violence*. Monroe, ME: Common Courage.

Farmer, Paul, Shirley Lindenbaum, and Mary-Jo Delvecchio Good. 1993. Women, poverty, and AIDS: An introduction. *Cultural Medical Psychiatry* 17 (4): 387–97.

Farquhar, Judith. 2002. *Appetites: Food and sex in post-Socialist China*. Durham, NC: Duke University Press.

Farquhar, Judith, and Qicheng Zhang. 2005. Biopolitical Beijing: Pleasure, sovereignty, and self-cultivation in China's capital. *Cultural Anthropology* 20 (3): 303–27.

Farrer, James. 2002. *Youth sex culture and market reform in Shanghai*. Chicago: University of Chicago Press.

Fausto-Sterling, Anne. 2000. *Sexing the body*. New York: Basic Books.

Feldman, Douglas A., ed. 1994. *Global AIDS policy*. Westport, CT: Bergin & Garvey.

Finnane, Antonia. 1996. What should Chinese women wear? *Modern China* 22 (2): 99–131.

Fisher, Jeffrey D., William A. Fisher, Stephen Misovich, Diane Kimble, and Thomas Malloy. 1992. Changing AIDS-risk behavior. *Psychological Bulletin* 111:453–74.

Fisher, Jeffrey D., and Stephen J. Misovich.1992. Impact of perceived social norms on adolescents' AIDS-risk behavior and prevention. In *Adolescents and AIDS: A generation in jeopardy*, ed. Ralph J. DiClemente, 117–36. Newbury Park: Sage.

Fitzgerald, John. 1996. *Awakening China, politics, culture, and class in the Nationalist Revolution*. Stanford, CA: Stanford University Press.

Flowers, Nancy. 1988. The spread of AIDS in rural Brazil. In *AIDS 1988: AAAS Symposia Papers*, ed. Ruth Kulstad, 159–73. Washington, DC: American Association Advance Science.

Foucault, Michel. 1978. *History of sexuality*. Vol. 1. New York: Random House.

Frankenberg, Ronnie. 1994. The impact of HIV/AIDS on concepts relating to risk and culture within British community epidemiology: Candidates or targets for prevention? *Social Science & Medicine* 38 (10): 325–35.

Freire, Paulo. 1970. *Pedagogy of the oppressed*. New York: Herder and Herder.

Furth, Charlotte. 1992. Chinese medicine and the anthropology of menstruation in contemporary Taiwan. *Medical Anthropology Quarterly* 6 (1): 27–48.

———. 1994. Rethinking Van Gulik: Sexuality and reproduction in traditional Chinese medicine. In *Endangering China:. Women, culture, and the state*, ed. Gail Hershatter, Christina K. Gilmartin, Lisa Rofel, and Tyrene White, ed. 125–46. Cambridge, MA: Harvard University Press.

Gao, Dewei. 2000. *Qingchun renge yu xingjiankang jiaoyu (Youth personality and sex health education)*. Beijing: Beijing Xingjiankang Jiaoyu Yanjiuhui (Beijing Sex Health Education Research Association).

Gao, Ersheng, Shaobo Xiao, Junqing Wu, and Wei Yuan. 2002. *Biyun jieyu youzhi fuwu yu zhiqing xuanze (Excellent contraceptive service and client-based choice)*. Beijing: Zhongguo Renkou Chubanshe.

Gao, Tian. 1919. Xing zhi sheng wu xue (Biology of sex). *Xin Qing Nian (New Youth)* 8 (6): 1–12.

Gao, Yaowu. 1998. Funu hunnei xingquanli xuyao falu baohu (Women's sex rights in marriage needs legal protection). *Women of China* 8:50–51.

Ge, Zihong. 2006. Shenyang shouli Aizibing ganranzhe shinianlai shenghuo Zhengchang (The first AIDS inflected in Shenyang has led a normal life for ten years). China.com.cn.

Geng, Xuebao. 1964. Yao hezuo, yao jianchi (Necessary cooperation and resolution). *Women of China* 9:26.

Giffin, Karen, and Catherine M. Lowndes. 1999. Gender, sexuality, and the prevention of sexually transmissible diseases: A Brazilian study of clinical practice. *Social Science & Medicine* 48:283–92.

Gil, Vincent E., Marco Wang, Allen F. Anderson, and Guao Matthew Lin. 1994. Plum blossoms and pheasants: Prostitutes, prostitution, and social control measures in contemporary China. *International Journal of Offender Therapy and Comparative Criminology* 38 (4): 319–37.

Gill, Bates, Jennifer Chang, and Sarah Palmer. 2002. China's HIV crisis. *Foreign Affairs*, March–April: 96–110.

Glick-Schiller, N. 1992. What's wrong with this picture? The hegemonic construction of culture in AIDS research in the United States. *Medical Anthropology Quarterly* 6 (3): 237–54.

Gomez, Cynthia A., and Barbara M. VanOss. 1996. Gender, culture, and power: Barriers to HIV-prevention strategies for women. *Journal of Sex Research* 33 (4): 355–62.

Goodkind, Daniel. 1997. Reasons for rising condom use in Vietnam. *International Family Planning Perspectives* 23:173–78.

Gordon, Linda, ed. 1979. *The struggle for reproductive freedom: Three stages of feminism.* New York: Monthly Review Press.

Gossett, Milton, and Jeremy Warshaw. 1992. The New York City campaign. In *AIDS: Prevention through education,* ed. Jaime Sepulveda et al., 283–96. New York: Oxford University Press.

Green, Edward, Janice A. Hogle, Vinand Nantulya, Rand Stoneburner, and John Stover. 2002. *What happened in Uganda? Declining HIV prevalence, behavior change, and the national response.* Washington, DC: U.S. Agency for International Development.

Greenhalgh, Susan, and Edwin A. Winckler. 2005. *Governing China's population.* Stanford, CT: Stanford University Press.

Gu, Sujuan. 1981. Tan jishu yuanyin (On technological reasons). *Women of China* 2:43.

Gu, Zhen. 1956a. Biyun yingxiang jiankang ma (Does contraception affect health). *Women of China* 7:26.

———. 1956b. Buyao suibian quzuo rengong liuchan (Don't do abortions casually). *Women of China* 6:24.

Guan, Shan. 2001. Buxiangxin Yanlei (I do not believe in tears). In *Yi lu ben zou (Marching on),* 18–143. Beijing: Huayi Chubanshe (Huayi Publishing House).

Guo, Huimin. 2002. Waiyu, ke raoshu de zui (Extramarital affairs are forgivable). *Zhongguo Funu (Women of China)* 8 (2): 28–29.

Gupta, Geeta Rao, and Ellen Weiss. 1993. Women's lives and sex: Implications for AIDS prevention. *Cultural Medical Psychiatry* 17 (4): 399–412.

Gutmann, Matthew. 2007. *Fixing men: Sex, birth control, and AIDS in Mexico.* Berkeley: University of California Press.

Guttmacher, Sally, et al. 1997. Condom availability in New York City Public high schools: relationships to condom use and sexual behavior. *American Journal of Public Health* 87:1427–33.

Hammar, Lydia. 1996. Bad canoes and bafalo: The political economy of sex on Daru Island, Western Province, Papua New Guinea. *Gender* 23:212–43.

Handwerker, Lisa. 1995. The hen that can't lay an egg: Conceptions of female infertility in modern China. In *Deviant bodies: Critical perspectives on difference in science and popular culture*, ed. Jennifer Terry and Jacqueline Urla. Bloomington: Indiana University Press.

Hanenberg, R. S., W. Rojanapithayakorn, P. Kunasol, and D. C. Sokal. 1994. Impact of Thailand's HIV-control program as indicated by the decline of sexually transmitted diseases. *Lancet* 344 (8917): 243–45.

Hansen, Helena, Maria Margarita Lopez-Iftikhar, and Margarita Alegria. 2002. The economy of risk and respect: Accounts by Puerto Rican sex workers of HIV risk taking. *Journal of Sex Research* 39 (4): 292–301.

Hao, Baiyu. 2005. Wen jiabao tichufangzhi aizibing de jiuxiang cuoshi (Wenjiabao proposed nine measures). *Dalian Daily*, June 16:A6.

Hart, G. J., R. Pool, G. Green, S. Harrison, S. Nyanzi, and J. A. Whitworth. 1999. Women's attitudes to condoms and female-controlled means of protection against HIV and AIDS in South-western Uganda. *AIDS Care* 11 (6): 687–98.

He, Chisheng. 1964. Zhe shi zisi zili de si xiang (This is a selfish thought). *Women of China* 8:30.

He, Mu. 2003. Waiyuzhong de Xingyinsu (The element of sex in extramarital affairs). *Jiatingshenghuo zhinan (Direction to Family Life)* 8 (218): 52–53.

He, Sanwei. 2005. Aizinusheng de qiqie zhuiwen (The sad questions of an AIDS female student). *Nanfang zhoumo (Southern Weekend)*, June 23:1.

Hearst, Norman, and Sanity Chen. 2004. Condom promotion for AIDS prevention in the developing world: Is it working? *Studies in Family Planning* 35 (1): 39–47.

Henrickson, Mark. 1990. A mobile HIV education, counseling, and testing unit: A pilot initiative. *AIDS Education Review* 2 (2): 137–44.

Henriot, Christian. 2001. *Prostitution and sexuality in Shanghai: A social history, 1849–1949*. Cambridge: Cambridge University Press.

Herdt, Gilbert, ed. 1996. *Third sex, third gender*. New York: Zone Books.

Herdt, Gilbert, and Andrew Boxer. 1991. Ethnographic issues in the study of AIDS. *Journal of Sexual Research* 28 (2): 171–87.

Herdt, Gilbert, and Shirley Lindenbaum. 1992a. Sexual identity and risk for AIDS among gay youth in Chicago. In *Sexual behavior and networking: Anthropological and socio-cultural studies on the transmission of HIV*, ed. Tim Dyson, 153–202. Liege: Derouaz-Ordina.

———, ed. 1992b. *The time of AIDS: Social analysis, theory, and method*. Newbury Park, CA: Sage.

Herdt, Gilbert, William L. Leap, and Melanie Sovine. 1991. Anthropology, sexuality and AIDS. *Journal of Sex Research* 28 (2): 167–69.

Hershatter, Gail. 1986. *The workers of Tianjin, 1900–1949*. Stanford, CA: Stanford University Press.

———. 1997. *Dangerous pleasures: Prostitution and modernity in twentieth-century Shanghai*. Berkeley: University of California Press.

Hershatter, Gail, and Emily Honig. 1988. *Personal voices: Chinese women in the 1980s.* Stanford, CA: Stanford University Press.

Hillier, Lynne, Lyn Harrison, and Deborah Warr. 1998. When you carry condoms all the boys think that you want it: Negotiating competing discourses about safe sex. *Journal of Adolescence* 21:15–29.

Himes, Norman E. 1963. *Medical history of contraception.* New York: Gamut Press.

Holland, Janet, Caroline Ramazanoglu, Sue Scott, Sue Sharpe, and Rachel Thomson. 1990. Sex, gender and power: Young women's sexuality in the shadow of AIDS. *Sociology of Health and Illness* 12:336–50.

———. 1991. Between embarrassment and trust: Young women and the diversity of condom use. In *AIDS: Responses, interventions, and care,* ed. Peter Aggelton, 127–48. London: Falmer Press.

———. 1992a. Pleasure, pressure, and power: Some contradictions of gendered sexuality. *Sociological Review* 40:645–74.

———. 1992b. Risk, power, and the possibility of pleasure: Young women and safer sex. *AIDS Care* 4 (3): 273–83.

———. 1994a. Achieving masculine sexuality: Young men's strategies for managing vulnerability. In *AIDS: Setting a feminist agenda,* ed. Tamsin Wilton. Southport: Taylor & Francis.

———. 1994b. Desire, risk, and control: The body as a site of contestation. In *AIDS: Setting a feminist agenda,* ed. Lesley Doyal and Tamsin Wilton, 61–79. Southport: Taylor & Francis.

Honig, Emily. 1986. *Sisters and strangers: Women in the Shanghai cotton mills, 1919–1949.* Stanford, CA: Stanford University Press.

Hsu, Mei-Ling, Wen-Chi Lin, and Tsui-Sung Wu. 2004. Representations of "Us" and "Others" in the AIDS news discourse: A Taiwanese experience. In *Sexual cultures in East Asia,* ed. Evelyne Micollier, 183–222. London: RoutledgeCurzon.

Hu, Xiaoyun. 2005. Xiaochu Aizibing qishi, luhai henyuan (A long road to eliminating prejudice against AIDS patient). *Jiankang wenzhai bao* (*Health Digest Newspaper*), May 10:1.

Huang, Shirlena, and Brenda S. A. Yeoh. 2008. Heterosexualities and the global(ising) city in Asia: Introduction. *Asian Studies Review* 32 (March): 1–6.

Hubbard, Philip. 2000. Desire/disgust: Mapping the moral contours of heterosexuality. *Progress in Human Geography* 24 (2): 191–17.

Hunt, Charles W. 1996. Social vs. biological: Theories on the transmission of AIDS in Africa. *Social Science & Medicine* 42 (9): 1283–96.

Hunter, Susan S. 2005. *AIDS in Asia: A continent in peril.* New York: Palgrave Macmillan.

Hyde, Sandra Teresa. 2007. *Eating spring rice: The cultural politics of AIDS in southwest China.* Berkeley: University of California Press.

Ibañez, Gladys E, Barbara Oss Marin, Cristina Villareal, and Cynthia Gomez. 2005. Condom use at last sex among unmarried Latino men: An event level analysis. *AIDS and Behavior* 9 (4): 433–41.

ICAF (International Council on Adolescent Fertility). 1989. Media as messengers: Shaping programs to entertain and educate. *Passages: International Council on Adolescent Fertility* 9 (1): 1–4.

Inciardi, James A., Hilary L. Surratt, and Paulo R. Telles. 2000. *Sex, drugs, and HIV/AIDS in Brazil.* Boulder, CO: Westview Press.

Ingham, Roger, Alison Woodcock, and Karen Stenner. 1991. Getting to know you . . . young people's knowledge of their partners at first intercourse. *Journal of Community and Applied Social Psychology* 1 (2): 117–32.

Jeffrey, Leslie Ann. 2002. *Sex and borders: Gender, national identity, and prostitution policy in Thailand.* Vancouver: UBC Press.

Jeffreys, Elaine. 2004. *China, sex, and prostitution.* London: RoutledgeCurzon.

———, ed. 2006. *Sex and sexuality in China.* Abingdon: Routledge.

Ji, Xiangde. 2005. Hunnei qiangjian lilun de lilun yuandian (The theoretical base of marital rape). *Zhongguo Faxuewang (Chinese Law Net)*, September 8.

Jian, Ping. 2001. Caifang shouji: Jingyan Dalian (Interview memoirs in Dalian). *Xinzhoukan (New Weekly)* 10:44.

Jiang, Deyuan, Shuquan Qu, Wei Liu, Kyung-Hee Choi, Rongjian Li, Deyuan Jiang, Yuejiao Zhou, et al. 2002. The potential for rapid sexual transmission of HIV in China: Sexually transmitted diseases and condom failure highly prevalent among female sex workers. *AIDS and Behavior* 6 (3): 267–75.

Jiang, Xiaoyuan. 2003. *Xing gan, Sex: Yizhong wenhua jieshi (Sexy, sex: One kind of cultural interpretations).* Haikou: Hainan chubanshe.

Jiang, Yunfei. 2005. Ezhi Aizi, lvxing chengnuo (Curbing AIDS infection, implementing promise). In *Dalian Ribao (Dalian Daily)*. Dalian.

Jiang, Zongtao. 1966. Ta zhenshi ge guanxin sheyuan de haoganbu (She is a good cadre concerned about other members). *Women of China* 1:25.

Jin, Gege. 2003. Baozhu zhangfu de yanmian (Protecting the face of the husband). *Zhongguo Funu (Women of China)* 10 (2): 37.

Jin, Ying. 2002. Anquantao guanggao (Condom Ads). *Xinwen Zhoukan (News Weekly)* November 27.

Jing, Feng. 2004. Fafang anquantao juefei "Fang AI" quanbu (Issuing condoms is by no means the complete AIDS prevention). *Xinhuawang (Xinhua Net)*, November 30.

Jing, Jun. 2006. The social origin of AIDS panics in China. In *AIDS and social policy in China*, ed. Arthur Kleinman, Joan Kaufman, and Tony Saich, 152–69. Cambridge, MA: Harvard University Asia Center.

Ju, Liya. 2006. *Diary of AIDS female university student.* Beijing: Beijing Publishing House.

Kaler, Amy. 2004. The moral lens of population control: Condoms and controversies in southern Malawi. *Studies in Family Planning* 35:105–15.

Kammerer, Cornelia Ann, Otome Klein Hutheesing, Ralana Maneeprasert, and Patricia V. Symonds. 1995. Vulnerability to HIV Infection among Three Hill Tribes in Northern Thailand. In *Culture and sexual risk*, ed. Han ten Brummelhuis and Gilbert H. Herdt, 53–75. Amsterdam: Gordon & Breach.

Kapumba, Sipo, V. Manda, and R. Zambezi. 1991. *Focus group research on condom use for AIDS prevention.* Ndola: Planned Parenthood Association of Zambia.

Katende, Charles, Ruth Knight, Reeru Gupta, Rodney Knight, and Cheryl Lettenmaier. 2000. *Uganda delivery of improved services for health evaluation surveys 1999.* Chapel Hill, NC: Measure Evaluation.

Kaufman, Joan, and Jun Jing. 2002. China and AIDS—the time to act is now. *Science* 296 (June 28): 2339–40.

Kaufman, Joan, Arthur Kleinman, and Tony Saich, ed. 2006a. *AIDS and Social Policy in China.* Cambridge, MA: Harvard University Asia Center.

———. 2006b. Introduction: AIDS and social policy in China. In *AIDS and social policy in China*, ed. Arthur Kleinman Joan Kaufman, and Tony Saich, 3–14. Cambridge, MA: Harvard University Asia Center.

Kaufman, Joan, and Kathrine Meyers. 2006. AIDS surveillance in China: Data gaps and research for AIDS policy. *In* AIDS and Social Policy in China. ed. Arthur Kleinman, Joan Kaufman, and Tony Saich. Cambridge, MA: Harvard University Asia Center.

Kegeles, Susan M., R. Greenblatt, J. Catania, C. Cardenas, J. Gottlieb, and T. Coates. 1989. AIDS risk behavior among sexually active Hispanic and white adolescent females. Fifth international conference on AIDS, Montreal.

Keller, Sarah N., and Jane Brown. 2002. Media interventions to promote responsible sexual behavior. *Journal of Sex Research* 39 (1): 67–72.

Kelly, J. A. 1992. AIDS prevention: Strategies that work. *The AIDS Reader*, July–August:135–41.

Kelly, Paula-Frances, ed. 2004. *What is known about gender, the constructs of sexuality, and dictates of behavior in Vietnam as a Confucian and socialist society and their impact on the risk of HIV/AIDS epidemic.* New York: RoutledgeCurzon.

Kerriga, Deanna, Jonathan Ellen, Luis Moreno, Santo Rosario, Joanne Katz, David Celentano, and Michael Sweat. 2003. Environmental-structural factors significantly associated with consistent condom use among female sex workers in the Dominican Republic. *AIDS* 17:415–23.

Kirby, Douglas, Nancy Brener, Nancy Brown, Nancy Peterfreund, Pamela Hillard, and Ron Harrist. 1998. The impact of condom distribution in Seattle schools on sexual behavior and condom use. *American Journal of Public Health* 89:182–87.

Klein, Megan. 2001. Social marketing and communications for health determinants of condom use among unmarried youth in Yaounde and Douala. PSI Research Division Working Paper No. 47. PSI (Population Services International). Washington, DC.

Klein, Megan, and Y. Coombes. 1999. *Trust and condom use: The role of sexual caution and sexual assurances for Tanzanian youth.* PSC Research Division Working Paper No. 64. PSI (Population Services International). Washington, DC.

Knodel, John, and Anthony Pramualratana. 1994. *Prospects for increased condom use in marital unions in Thailand.* PSC Research Report No. 95-337. PSC Research Division Working Paper No. 64. PSI (Population Services International). Washington, DC.

Ko, Dorothy. 1994. *Teachers of the inner chambers: Women and culture in seventeenth-century China*. Stanford, CA: Stanford University Press.

Koetswang, Suporn, and N. J. Ford. 1999. *A self-esteem and personal future-focused intervention programme to promote condom use by female sex workers in Thailand*. Nakhon Pathom, Thailand: Institute for Population and Social Research, Mahidol University.

Ku, Andrzej. 2004. The sociocultural context of condom use within marriage in rural Lebanon. *Family Planning* 35141:246–60.

Kuang, Caiwei. 1979. Wei sihua zhisheng yige zinu guangrong (It is glorious to give birth to one child for four modernizations). *Women of China* 8:24.

Lamptey, Peter, and Gail A. W. Goodridge. 1991. Condom issues in AIDS prevention in Africa. *AIDS* 5 (Suppl. 1): S183–S191.

Lan, Huaisi. 2006. Directing a perfect sex (daoyan yichang wanmei xingai). *Hers* 2 (20): 182–84.

Lane, Sandra D. 1997. Television minidramas: Social marketing and evaluation in Egypt. *Medical Anthropology Quarterly* 11 (2): 164–82.

Lang, Jinghe. 1980. Yousheng de diyibu—hunqian jiancha (The first step in eugenics: Premarital test). *Women of China* 8:46–47.

Lao, Yu. 2005. Tantao jinji biyunyao de daodewenti (On the moral issues of emergency contraceptives). *Xin zhoukan* (*New Weekly*), November 2.

Laqueur, Thomas. 1990. *Making sex: Body and gender from the Greeks to Freud*. Cambridge, MA: Harvard University Press.

Larson, Wendy. 2002. The self-loving the self: Men and connoisseurship in modern Chinese literature. In *Chinese femininities, Chinese masculinities*, ed. Jeffrey Wasserstrom and Susan Brownell, 175–94. Berkeley: University of California Press.

Lau, Joseph I. E., A. S. Tang, and H. Y. Tsui. 2003. The relationship between condom use, sexually transmitted diseases, and location of commercial sex transaction among male Hong Kong clients. *AIDS* 17:105–12.

Law, Lisa. 2000. *Sex work in Southeast Asia: The place of desire in a time of AIDS*. London: Routledge.

Lear, Dana. 1995. Sexual communication in the age of AIDS: The construction of risk and trust among young adults. *Social Science & Medicine* 41:1311–23.

Leclerc-Madlala, Suzanne. 2001. Virginity testing: Managing sexuality in a maturing HIV/AIDS epidemic. *Medical Anthropology Quarterly* 15:533–52.

Lee, Ching Kwan. 1998. *Gender and the South China miracle*. Berkeley: University of California Press.

Lei, Jieqiong. 1957. He nianqingren tan hunshi (Talking about marriage with young people). *Women of China* 4:24–25.

Lei, Zhenwu. 1979. Shouren kuazai de jieyu cuoshi (Applaudable birth control measures). *Women of China* 11:45.

Lewin, Kurt. 1958. The group reason and social change. In *Readings in social psychology*, ed. Eleanor Maccoby. London: Holt, Rinehart and Winston.

Li, Fei. 1958. *Xingzhishi wenda* (*Questions and answers of sex knowledge*). Baoding: Hebei Renmin Chubanshe.

Li, Jinxing. 2000. Jinanshi shouli HIV-1 ganranzhe diaochabaogao (Report of HIV-1 infected people in Jinan). *Zhongguo Xingbing Aizibing Fangzhi (Prevention of STI and AIDS in China)* 6 (2): 113.

Li, Qingshan. 2005. Guilin gongshang (Guilin industrial commerce). *Zhongguo Xiaofeizhe Bao (Chinese Consumer Newspaper)*, July 25.

Li, Shunlai. 2000. Anquantao Guanggao Women Zai Dengdai (We are Waiting for Condom Ads). Zhongguo Funu bao (Chinese Women Newspaper) April. 11.

Li, Tong. 2001. Falu gaibugai gei anquantao yige mingfen? (Should the law give condom a name?). *Beijing Qingnian Bao (Beijing Youth Newspaper)*, September 7.

Li, Xiaofeng, Li Ma, Xiaohong Gao, Cuili Zhang, Shu Zhang, and Chengzhi Lv. 2004. 1999, 2000 Nian Dalian shi xingbing liuxingbingxue fenxi (An epidemiological analysis of STIs in Dalian in 1999 and 2000). *Shiyong Yufang Yixue (Pragmatic Prevention Medicine)* 3:16–17.

Li, Xiongqiong. 1957. Woneng daying tade yaoqiu ma (Can I say yes to his request?). *Women of China* 3:14–15.

Li, Yali. 2002. Bamitude zhangfu daihuijia (Bring back the astraying husband). *Zhongguo Funu (Women of China)* 8 (2): 26–28.

Li, Yaling. 1999. Kaojin mingpai daxue liangcaizi bei xiaojie yinyou ranshang linbing (An outstanding student in a key university is seduced by a hostess and contracted a STD). *In Chengdu Shangbao (Chengdu Commerce Newspaper)*, 2.

Li, Yanchun. 2004. Anquantao anweibuliao posuidexin (Condom cannot comfort a broken heart). *Beijing Qingnianbao (Beijing Youth Newspaper)*, November 29.

Li, Yinhe. 2002. Anquantao lunzheng de beihou (Behind the debate of condoms). 21 shiji huanqiu baodao (Report around the world in the 21st century), December 26.

Li, Zengqing. 2003. Daxuesheng hunqian xingxingwei, tongju xianxiang sikao (Thoughts on cohabitation and sex conducts of college students). *Jiatingbaojian (Family Health)* 12:22–23.

Li, Zhongfeng. 2004. Wuyouwulu de ai diaocha: Anquantao Guanggao (Love without worries: Condom ads). *Shichang Bao (Market Newspaper)*, July 11.

Li, Zhongmao. 2005. Zhongguo de sici xinggeming (Four sex revolutions in China). *Wenzhai Zhoubao (Digest Weekly Newspaper)*, June 7:1.

Liang, Jun, and Kongling Xu. 1997. Jihuashengyu yu funushengyujiankang zhi libi (The drawback and benefit of family planning to women). In *Shengyu de Chuantong yu Xiandaihua (The tradition of reproduction and modernization)*, ed. Xiaojiang Li, 42–54. Zhengzhou: Henan Renmin Chubanshe.

Liang, Qiusheng, and Che-Fu Lee. 2006. Fertility and population policy: An overview. In *Fertility, family planning, and population policy in China*, ed. Dudley L. Poston, Chiung-Fang Chang, Che-Fu Lee, Sherry L. McKibben, and Carol S. Walther, 8–20. London: Routledge.

Liang, Yong. 2003. Anquantao guanggao shang yangshi shiguanniande jinbu (It is ideological progress for condom ads to be on CCTV). *Xinhuawang (Xinhua Net)*, November 28.

Liang, Zhao. 1957. Jihua shengyu bing bunan (It is not difficult to have family planning). *Women of China*, no. 4:25.

Lin, Ho Swee. 2008. Private love in public space: love hotels and the transformation of intimacy in contemporary Japan. *Asian Studies Review* 32 (March): 31–56.

Lin, Minduo. 2005. Ezhi gaoweirenqun aizibing chuanbo de sikao (Reflections on blocking HIV/AIDS high-risk population). In *Zhongguo Aizibing fangzhi (Prevention of HIV/AIDS in China)*. Beijing: Capital University of Medical Science.

Lin, Peng, and Hua Wang. 2000. Guangdongsheng HIV/AIDS jiancexitong de jianli jiqi xiaoguopingjia (Guangdong HIV/AIDS surveillance system). *Zhongguo Xingbing Aizibing Fangzhi (Prevention of STI and AIDS in China)* 6 (3): 133–34.

Lin, Quanyi. 2002. Training on condom social marketing conducted in Urumqi, Xinjiang. International Conference on AIDS, Manchester, July 7–12.

Lin, Wei. 1998. Biyuntao nengzuo guanggao ma? (Can condoms be advertised?). *Zhongguo Qingnian Bao (China Youth Daily)*, December 4.

Lindan, Christina, S. Allen, M. Carael, F. Nsengumuremyi, P. Van de Perre, A. Serufilira, J. Tice, D. Black, et al. 1991. Knowledge, attitudes, and perceived risk of AIDS among urban Rwandan women: Relationship to HIV infection and behavior change. *AIDS* 5 (8): 993–1002.

Lindenbaum, Shirley. 1997. AIDS: Body, mind, and history. In *AIDS in Africa and the Caribbean*, ed. George C. Bond, John Kreniske, Ida Susser, and Joan Vincent, 191–94. Boulder, CO: Westview.

———. 1998. Images of catastrophe: The making of an epidemic. In *The political economy of AIDS*, ed. Merrill Singer, 33–58. Amityville, NY: Baywood.

Ling, Feng. 2004. Wuzhi jiaju "Kongai Zheng" (Ignorance exacerbate paranoid of AIDS). *Jiankang bao (Health Newspaper)*, November 29:1.

Ling, Yun. 2003. Gaozhi fuqi tongshuo hunwaiqing (Speaking of extramarital affairs). Zhongguo Funu (*Women of China*) 1 (2): 34.

Little, J. 2003. Riding the rural love train: Heterosexuality and the rural community. *Sociologia Ruralis* 43 (4): 401–17.

Liu, Dalin, et al. 1997. *Sexual behaviour in modern China*. New York: The Continuum.

Liu, Huizhen. 1966. Shishuo buru yizuo (Ten speeches speak lower than one deed). *Women of China* 1:24.

Liu, Jianlin. 1999. Sanpeinu chengwei fanzui gaofa qunti (The highest crime rate is found in the group of bar hostesses). In *Zhongguo Qingnianbao (Chinese Youth Newspaper)*, 3.

Liu, Jianqiang. 2005 Zhongyang dangxiaoli de aizi guannian jiaofeng (Conflict of views on AIDS in central party schools). *Nanfang Zhoumo (South Weekend)*, June 23:A6.

Liu, Jiansheng. 2004. Qieshijiaqiang aizibing fangzhi gongzuo (Strengthening AIDS prevention work). *Fazhi Ribao (Law Daily)*, December 1:1.

Liu, Wei, and Zhou Yuejiao. 2001. Evaluation of an intervention designed to reduce risk behaviors among woman engaged in illegal commercial sex activities in frontier areas of Guangxi province. *Chinese Journal of STD/AIDS Prevention and Control* 4:223–25.

Liu, Xingyu, and Jiang Xu. 2005. Hunyin: Nurende quanli he zeren (Marriage: Women's rights and responsibilities). *Zhongguo Funu* (*Women of Women*), April 2 (647): 32–33.

Liu, Yuanli, and Joan Kaufman. 2006. Controlling HIV/AIDS in China: Health system challenges. In *AIDS and social policy in China*, ed. Arthur Kleinman, Joan Kaufman, and Tony Saich, 75–95. Cambridge, MA: Harvard University Asia Center.

Liu, Zhen. 2002. Feifatongju: Weixiandeshishang (Illegal cohabitation: dangerous style). *Zhongguo Funu* (*Women of China*) 6:18.

Liu, Zheng, and Cangping Wu. 1979. Funumen, zhengdang jihua shengyu gongzuo de cujin pai (Women, fight as the harbinger of family planning). *Women of China* 8:22–23.

Liu, Ziliang. 2004. Zhimian Aizi (Confronting AIDS). In *Mianduimian* (*Face to Face*), ed. Jianzeng Liang, 149–62. Changchun: Jilin Renmin Chubanshe.

Longfield, Kim, Megan Klein, and Joh Berman. 2002. *Criteria for trust and how trust affects sexual decision-making among youth*. Washington, DC: Population Services International.

Luo, Gang. 2005a. E zhu AIDS manyan de yanhou (Strangle the throat of AIDS transmission). *Jiankang Wenzhai Bao* (*Health and Digest Newspaper*), no. 709 (March 20): 8.

———. 2005b. Wo jiu zai ni shenpao (I am right by your side). *Jiankang Bao* (*Health Newspaper*), April 20:7.

Lupton, Deborah. 1994. *Moral themes and dangerous desires: AIDS in the news media*. London: Taylor & Francis.

Lyttleton, Chris. 1996. Health and development: Knowledge systems and local practice in rural Thailand. *Health Transition Review* 6 (1): 25–48.

Ma, Jinyu. 2005. Yuaizibing nudaxuesheng mianduimian: Zai aiqingzhong shalu qingchun (Facing an AIDS female university student: Killing the youth in love). *Nanfang Renwuzhoukan* (*South Weekly*), June 3:4.

Ma, Qinghuai. 1954. Lenin talks about the issue of women, marriage and sex. *Xin Zhongguo Funu* (*New Women of China*) 51 (1): 6–7.

Macaluso, M., M. J. Demand, L. M. Artz, and E. W. Hook. 2000. Partner type and condom use. *AIDS* 14:537–46.

MacPhail, Catherine, and Catherine Campbell. 2001. I think condoms are good but, aai, I hate those things; condom use among adolescents and young people in a Southern African township. *Social Science & Medicine* 52:1613–27.

Maharaj, Pranitha. 2001. Obstacles to negotiating dual protection: Perspectives of men and women. *African Journal of Reproductive Health* 5 (3): 150–61.

———. 2004. Perception of risk of HIV infection in marital and cohabiting partnerships. *African Journal of AIDS Research* 3 (2): 131–43.

Maharaj, Pranitha, and John Cleland. 2004. Condom use within marital and cohabitating partnerships in Kwa-Zulu-Natal, South Africa. *Studies in Family Planning* 35 (2): 116–24.

Mahler, K. 1996 Condom use increase in Norway appears related more to contraception than to disease prevention. *Family Planning Perspectives* 28 (2): 82–83.

Mann, Jonathan. 1996. *Human rights and AIDS: The future of the pandemic.* New York: Plenum.

Mann, Susan. 1997. *Precious records: Women in China's long eighteenth century.* Stanford, CA: Stanford University Press.

Marandu, Edward E., and Mbaki A. Chamme. 2004. Attitudes towards condom use for prevention of HIV infection in Botswana. *Social Behavior and Personality* 32 (5): 491–510.

Martin, Barbara Van Oss, and Cynthia A. Gomez. 1996. Latino culture and sex: Implications for HIV Prevention. In *Psychological interventions and research with Latino populations*, ed. Jorge G. Garcia Maria Cecilia Zea, 73–93. Meedham Heights, MA: Allyn and Bacon.

Maticka-Tyndale, Eleanor. 1991. Sexual scripts and AIDS prevention: Variations in adherence to safer-sex guidelines by heterosexual adolescents. *Journal of Sex Research* 28:45–66.

———. 1992. Social construction of HIV transmission and prevention among heterosexual young adults. *Social Problems* 39:238–52.

Maticka-Tyndale, Eleanor, David Elkins, Melissa Haswell-Elkins, Darunee Rujkarakorn, Thicumporn Kuyyakanond, and Kathryn Stam. 1997. Contexts and patterns of men's commercial sexual partnerships in northeastern Thailand: Implications for AIDS prevention. *Social Science & Medicine* 44 (2): 199–213.

Matuszak, Sascha. 2003. Safe sex in China. Antiwar.com, June 13.

McGrath, Janet. 1993. Anthropology and AIDS. *Social Science & Medicine* 36 (4): 429–39.

McMahon, James M., Stephanie Tortu, Enrique R. Pouget, Rahul Hamid, and Alan Neaigus. 2006. Contextual determinants of condom use among female sex exchanges in East Harlem, NYC: An event analysis. *AIDS and Behavior* June 16:731–41.

McMillan, Jo. 2006. Selling sexual health: China's emerging sex shop industry. In *Sex and sexuality in China*, ed. Elaine Jeffreys, 124–38. London: Routledge.

Meekers, Dominique. 1999. Patterns of use of the female condom in urban Zimbabwe. PSI Research Division Working Paper No. 28. PSI (Population Services International). Washington, DC.

Mehryar, Amir. 1995. Condoms: Awareness, attitudes, and use. In *Sexual behavior and AIDS in the developing world*, ed. John Cleland and Benoit Ferry, 124–56. London: Taylor & Francis.

Meng, Tie. 2006. Yeti anquantao neng shamie bingdu (Liquid condoms can kill viruses). *Hangzhou Ribao (Hangzhou Daily)*, August 9.

Messersmith, Lisa J., Thomas T. Kane, Adetanwa I. Odebiyi, and Alfred A. Adewuyi. 2000. Who's at risk? Men's STD experience and condom use in southwest Nigeria. *Studies in Family Planning* 31 (September): 203–16.

Micollier, Evelyne. 2003. HIV/AIDS-related stigmatization in Chinese society: Bridging the gap between official responses and civil society. A cultural approach to HIV/AIDS prevention and care. UNESCO/UNAIDS Research Project No. 20. Paris, France: UNESCO.

———, ed. 2004a. *Sexual cultures in East Asia.* New York: RoutledgeCurzon.

————, ed. 2004b. Social significance of commercial sex work: Implicitly shaping a sexual culture? In *Sexual cultures in East Asia*, ed. Evelyne Micollier, 1–23. New York: RoutledgeCurzon.

————. 2005a. AIDS in China: Discourses on sexuality and sexual practices. *China Perspectives* 60: 2–14.

————. 2005b. Collective mobilisation and transnational solidarity to combat AIDS in China: Local dynamics and visibility of groups sexual and social minorities. *Perspectives on Health* 7:30–38.

————. 2006. Sexualities and HIV/AIDS vulnerability in China, an anthropological perspective. *Sexologies* 15:192–201.

Middel, Anne. 2001. Interpretations of condom use and nonuse among young Norwegian gay men: A qualitative study. *Medical Anthropology Quarterly* 15 (1): 58–83.

Miller, Maureen and Alan Neaigus. 2002. An economy of risk: Resource acquisition strategies of inner city women who use drugs. *International Journal of Drug Policy* 13 (5): 409–18.

Mills, Singh, P. Benjarattanaporn, A. Bennett, R. N. Pattalung, D. Sundhagul, P. Trongsawad, S. E. Gregorich, N. Hearst, and J. S. Mandel. 1997. HIV risk behavioral surveillance in Bangkok, Thailand: Sexual behavior trends among eight population groups. *AIDS* 11 (Suppl. 1): S43–S51.

Mnyinka, K. S., G Kvale, and K. I. Klepp. 1995. Perceived function of and barriers to condom use in Arusha and Kilimanjaro regions of Tanzania. *AIDS Care* 7 (3): 295–305.

Mo, Yunshi. 1964. Xiading juexin, yangcheng xiguan (Make a determination, form a habit). *Women of China* 5 (31).

Molm, Linda, Nobuyuki Takahashi, and Gretchen Peterson. 2000. Risk and trust in social exchange: An experimental test of a classical proposition. *American Journal of Sociology* 105:1396–1427.

Moore, Susan, & Doreen Rosenthal. 1992. The social context of adolescent sexuality: Safe sex implications. *Journal of Adolescence* 15:415–35.

Morris, Martina, Anthony Pramualratana, Chai Podhisita, and M. J. Wawer. 1995. The relational determinants of condom use with commercial sex partners in Thailand. *AIDS* 9 (5): 507–15.

Mosse, George L. 1985. *Nationalism and sexuality*. Madison: University of Wisconsin Press.

Mukasa, Rebecca, Janet W. McGrath, Charles B. Rwabukwali, Debra A. Schumann, Jonnie Pearson-Marks, Barbara Namande, Sylvia Nakayiwa, and Lucy Nakyobe. 1992. Cultural determinants of sexual risk behavior for AIDS among Baganda women. *Medical Anthropology Quarterly* 6 (2): 153–61.

Ngai, Pun. 2005. *Made in China: Women factory workers in a global workplace*. Durham, NC: Duke University Press.

Nie, Jing-bao. 2005. Behind the silence: Chinese voices on abortion. New York: Rowman & Littlefield.

Nie, Mao. 2004. *Aifengle, lefengle (Love is crazy)*. Beijing: Minzu Chubanshe.

Obbo, Christine. 1995 Gender, age, and class: Discourses on HIV transmission and control in Uganda. In *Culture and sexual risk: Anthropological perspectives on AIDS*, ed. Han ten Brummelhuis and Gilbert Herdt, 79–95. Amsterdam: Gordon Breach.

Oladosu, Muyiwa. 2001. Consistent condom use among sex workers in Nigeria. PSI Research Division Working Paper No. 39. PSI (Population Services International). Washington, DC.

O'Leary, Ann. 2000. Women at risk for HIV from a primary partner: Balancing risk and intimacy. *Annual Review of Sex Research* 11:191–234.

Pan, Jianqing. 2005. Qiantan "hunnei qiangjian" (Discussion of "marital rape"). *Jiaxing Sifa Xingzhengwang (Jianxing Law Executive)* 11:12–30.

Pan, Mingxin. 2004. Anquan tao-anquan fangbian kuaile (Condom—safe, convenient, and happy). *Shenghuo yu Jiankang (Life and Health)* 2:51.

Pan, Suiming. 2006. Transformation in the primary life cycle: The origins and nature of China's sexual revolution. In *Sex and sexuality in China*, ed. Elaine Jeffreys, 21–42. London: Routledge.

———. 2007. HIV and FSWs. Seminar on HIV prevention and sex work, Beijing, UNDP. June 4.

Parish, William L., Edward O. Laumann, Myron S. Cohen, Suiming Pan, Heyi Zheng, Irving Hoffman, Tianfu Wang, and Kwai Hang Ng. 2003. Population-based study of chlamydia infection in China: A hidden epidemic. *JAMA* 289 (10): 1303–5.

Parish, William L., and Suiming Pan. 2006. Sexual partners in China: Risk pattern for infection by HIV and possible interventions. In *AIDS and social policy in China*, ed. Arthur Kleinman, Joan Kaufman, and Tony Saich, 190–213. Cambridge, MA: Harvard University Asia Center.

Parker, Andrew, Mary Russo, Doris Sommer, and Patricia Yaeger. 1992. Introduction. In *Nationalisms and sexualities*, ed. Andrew Parker, Mary Russo, Doris Sommer, and Patricia Yaeger, 1–18. New York: Routledge.

Parker, Richard G. 1987. Acquired Immunodeficiency Syndrome in urban Brazil. *Medical Anthropology Quarterly*, new series, 1:155–72.

———. 1988. Sexual culture and AIDS education in urban Brazil. In *AIDS 1988: AAAS symposia papers*, ed. Ruth Kulstad, 269–89. Washington, DC: American Association Advance Science.

———. 1991. *Bodies, pleasures, and passions: Sexual culture in contemporary Brazil*. Boston: Beacon Press.

———. 1994. Sexual cultures, HIV transmission, and AIDS prevention. *AIDS* 8 (Suppl.): S309–S314.

Parker, Richard G., Delia Easton, Charles H. Klein. 2000. Structural barriers and facilitators in HIV prevention: A review of international research. *AIDS* 14 (Suppl. 1): S22–S32.

Parker, Richard. 2001. Sexuality, culture, and power in HIV/AIDS research. *Annual Review of Anthropology* 30: 163–79.

Paxson, Heather. 2002. Rationalizing sex: Family planning and the making of modern lovers in urban Greece. *American Ethnologist* 29 (2): 307–34.

Perkins, Roberta, and Garrett Prestage. 1994. *Sex work and sex workers in Australia.* Sydney: University of New South Wales Press.

Pfeiffer, James. 2004. Condom social marketing, Pentecostalism, and structural adjustment in Mozambique: A clash of AIDS prevention messages. *Medical Anthropology Quarterly* 18 (1): 77–103.

Phoolcharoen, Wiput. 1998. HIV/AIDS prevention in Thailand: Success and challenges. *Science* 280 (5371): 1873–74.

Pigg, Stacy Leigh, and Simon Fraser. 2001. Languages of sex and AIDS in Nepal: Notes on the social production of commensurability. *Cultural Anthropology* 16 (4): 491–541.

Piot, Peter, and Marie Laga. 1991. Current approaches to sexually transmitted disease control in developing countries. In *Research issues in human behavior and sexually transmitted diseases in the AIDS era*, ed. Judith N. Wasserheit, Sevgi O. Aral, King K. Holmes, and Penelope J. Hitchcock, 281–95. Washington, DC: American Society for Microbiology.

Plant, Michael A. 1993. *AIDS, drugs, and prostitution.* London: Routledge.

Plummer, Mary L., Daniel Wight, Joyce Wamoyi, Gerry Mshana, Richard J. Hayes, and David A. Ross. 2006. Farming with your hoe in a sack: Condom attitudes, access, and use in rural Tanzania. *Studies in Family Planning* 37 (1): 29–40.

Porter, Doug. 1997. A plague on the borders: HIV, development, and traveling identities in the Golden Triangle. In *Sites of desire: Economies of pleasure*, ed. Lenore Manderson and Margaret Jolly, 212–32. Chicago: University of Chicago Press.

Porter, Robert W. 1994. AIDS in Ghana: Priorities and policies. In *Global AIDS policy*, ed. Douglas A. Feldman, 90–106. Westport, CT: Bergin & Garvey.

Preston-Whyte, Eleanor, and Maria Zondi. 1991. Adolescent sexuality and its implications for teenage pregnancy and AIDS. *Continuing Medical Education* 9 (11): 1389–94.

Pritchard, Annette, and Nigel Morgan. 2006. Hotel Babylon? Exploring hotels as liminal sites of transition and transgression. *Tourism Management* 27 (5): 762–72.

Pyett, Priscilla M., and Deborah J. Warr. 1997. Vulnerability on the streets: Female sex workers and HIV risk. *AIDS Care* 9:539–47.

Qu, Shuquan. 1997. 1996 nian zhongguo aizibing shaodian jiance baogao (Report of sentinel surveillance in China in 1996). *Zhongguo Xingbing Aizibing Fangzhi (Prevention of STI and AIDS in China)* 3 (5): 193–98.

———. 2001. National AIDS/HIV surveillance and current HIV epidemic status in China. National Workshop on HIV/STI Surveillance, Nanjing, China.

Qu, Shuquan, Wei Liu, Kyung-Hee Choi, Rongjian Li, Deyuan Jiang, Yuejiao Zhou, Fang Tian, et al. 2002. The potential for rapid sexual transmission of HIV in China: Sexually Transmitted Diseases and Condom Failure Highly Prevalent among Female Sex Workers. *AIDS and Behavior* 6 (3).

Qu, Xuesong. 2005. Moba xingjiaoyu biancheng xingshanghai (Don't turn sex education to sexual abuse). *Zhongguo Funu (Women of China)*, no. 2 (October): 46.

Ramazanoglu, Caroline, and Janet Holland. 1993 Women's sexuality and men's appropriation of desire. In *Up against Foucault: Explorations of some tensions between Foucault and feminism*, ed. Caroline Ramazanoglu, 239–64. London: Routledge.

Reed, Dave, and Martin S. Weinberg. 1984. Premarital coitus: Developing and establishing sexual scripts. *Social Psychology Quarterly* 47:129–38.

Ren, Min. 2003. Yeti biyuntao queyou qishi (True liquid condoms). *Nanfang Dushi Bao (South City Newspaper)*, February 3.

Ren, Qingyun. 1997. Yingxiang zhongyuan nongcun funu shengyu jiankang de ruogan yinsu (Elements that affect rural women's reproductive health). In *Shengyu: Chuantong yu xiandaihua (Reproduction: Tradition and Modernity)*, ed. Xiaojiang Li, 153–54. Zhengzhou: Henan Renmin Chubanshe.

Ren, Xin. 1999. Prostitution and economic modernization in China. *Violence against Women* 5 (12): 1411–36.

Renaud, Michelle Lewis. 1997. *Women at the crossroads: A prostitute community's response to AIDS in urban Senegal.* New York: Routledge.

Roche, Brenda, Alan Neaigus, and Maureen Miller. 2005. Street smarts and urban myths: Women, sex work, and the role of storytelling in risk reduction and rationalization. *Medical Anthropology Quarterly* 19 (2): 149–70.

Rofel, Lisa. 2007. Desiring China: Experiments in neoliberalism, sexuality, and public culture. Durham, NC: Duke University Press.

Rogers, Everett, S. Aikat, S. Chang, P. Poppe, and P. Sopory. 1989. Proceedings from the Conference on Entertainment-Education for Social Change. Annenberg School of Communications and the University of Southern California, Los Angeles.

Rojanapithayakorn, Wiwat, and R. Hanenberg. 1996. The 100% condom program in Thailand. *AIDS* 10:1–7.

Rong, Dongyue. 1999. Anquantao daodi zenmo xuanchuan. *Beijing Wanbao (Beijing Evening Newspaper)*, November 30.

Rossem, Ronan Van, Dominique Meekers, and Zacch Akinyemi. 2000. Condom use in Nigeria: Evidence from two waves of a sexual behavior and condom use survey. PSI Research Division Working Paper No. 31. PSI (Population Services International). Washington, DC.

Rou, Keming, Zunyou Wu, Sheena G. Sullivan, Fan Li, Jihui Guan, Jihui Xu, Dahua Liu, et al. 2007. A five-city trial of a behavioural intervention to reduce sexually transmitted disease/HIV risk among sex workers in China. *AIDS* 21 (8): 95–101.

Rowe, William T. 1984. *Hankow: Commerce and society in a Chinese city: 1796–1889.* Stanford, CA: Stanford University Press.

Ru, Xiaomei. 2006. Youth and HIV/AIDS in China. In *AIDS and social policy in China*, ed. Arthur Kleinman, Joan Kaufman, and Tony Saich, 232–42. Cambridge, MA: Harvard University Asia Center.

Ruan, Fangyu. 1991. *Sex in China: Studies in sexology in Chinese culture.* New York: Plenum.

Rwabukwali, Charles B., Janet McGrath, and Debra Schumann. 1991. Socioeconomic determinants of sexual risk behavior among Baganda women in Kampala, Uganda. Seventh International Conference on AIDS, Florence, Italy. June 16–21.

Sacks, Valerie. 1996. Women and AIDS: An analysis of media misrepresentations. *Social Science & Medicine* 42 (1): 59–73.

Saich, Tony. 2006. Social policy development in the era of economic reform. In *AIDS and social policy in China*, ed. Arthur Kleinman, Joan Kaufman, and Tony Saich, 15–46. Cambridge, MA: Harvard University Asia Center.

Sanders, Teela. 2004. The risks of street prostitution: Punters, police, and protesters. *Urban Studies* 41 (9): 1703–17.

Schoepf, Brooke. 1988. Women, AIDS, and economic crisis in Central Africa. *Canadian Journal of African Studies* 22 (3): 625–44.

———. 1991. Ethical, methodological, and political issues of AIDS research in Central Africa. *Social Science & Medicine* 33:749–63.

———. 1992. AIDS, sex, and condoms: African healers and the reinvention of tradition in Zaire. *Medical Anthropology* 14:225–42.

———. 1995. Culture, sex research, and aids prevention in Africa. In *Culture and sexual risk: anthropological perspectives on AIDS*, ed. Han ten Brummelhuis and Gilbert Herdt, 29–51. Amsterdam: Gordon Breach.

———. 2001. International AIDS research in anthropology: Taking a critical perspective on the crisis. *Annual Reviews of Anthropology* 30:335–61.

Schoepf, Brooke G. 1998. Inscribing the body politic: Women and AIDS in Africa. In *Women and biopower: What constitutes resistance?* ed. Patricia Kaufert and Margaret Lock, 98–126. New York: Cambridge University Press.

Schoepf, Brooke G., Rukarangira-Wa Nkera, Claude Schoepf, Engundu Walu, and Payanzo Ntsomo. 1988. AIDS and society in Central Africa: A view from Zaire. In *AIDS in Africa: Social and policy impact*, ed. Richard C. Rockwell and Norman Miller, 211–35. Lewiston, ME: The Edwin Mellen Press.

Schoepf, Brooke G., Claude Schoepf, and Joyce V. Milien. 2000. Theoretical therapies, remote remedies: SAPS and the political ecology of health in Africa. In *Dying for growth: Global inequality and the health of the poor*. Jim Y. Kim and Joyce V. Millen, 91–125. Monroe, ME: Common Courage Press.

Schumann, Debra A, Janet W. McGrath, Charles B. Rwabukwali, Jonnie Pearson-Marks, Barbara Namande, Lucy Nakyobe, Sylvia Nakayiwa, and Rebecca Mukasa. 1991. Culture and the risk of AIDS: The social organization of sexual risk behavior in Uganda. American Association Meetings, Chicago. November 20–24.

Scott, James. 1976. *The moral economy of the peasant: Rebellion and subsistence in Southeast Asia*. New Haven, CT: Yale University.

Seeley, Janet A., Sam S. Malamba, Andrew J. Nunn, Daan W. Mulder, Jan F. Kengeya-Kayondo, and Thomas G. Barton. 1994. Socioeconomic status, gender, and risk of HIV-1 infection in a rural community in south west Uganda. *Medical Anthropology Quarterly* 8 (1): 78–89.

Seidel, Gill. 1993. The competing discourses of HIV/AIDS in sub-Sahara Africa: Discourses of rights and empowerment vs. discourses of control and exclusion. *Social Science & Medicine* 36 (3): 175–94.

Sellers, Deborah, Sarah A. McGraw, and John B. McKinlay. 1994. Does the promotion and distribution of condoms increase teen sexual activity? Evidence from an HIV prevention program for Latino Youth. *American Journal of Public Health* 84:1952–58.

Serovich, Julianne, and Kathryn Greene. 1997. Predictors of adolescent sexual risk taking behaviors which put them at risk for contracting HIV. *Journal of Youth and Adolescence* 26:429–44.

Setel, Philip W. 1999. *A plague of paradoxes: AIDS, culture, and demography in northern Tanzania.* Chicago: University of Chicago Press.

Shao, Jing. 2006. Fluid labor and blood money: The economy of HIV/AIDS in rural central China. *Cultural Anthropology* 21 (4): 535–69.

Shen, Hsiu-Hua. 2008. The purchase of transnational intimacy: Women's bodies, transnational masculine privileges in Chinese economic zones. *Asian Studies Review* 32 (March): 57–75.

Shen, Jie, Kangmai Liu, Mengjie Han, and Fujie Zhang. 2004. The China HIV/AIDS epidemic and current response. In *AIDS in Asia: The Challenge Continues*, ed. Jai P. Narain, 171–90. California: Sage.

Shen, Wei. 2001. Gonggong zaiqu e'xi shi ci mousha momo (The malicious second wife attempts to murder her man's wife ten times). In *Xin Shang Bao* (*New Commerce Newspaper*), 20. Dalian Newspaper Conglomerate.

Shen, Ying. 2007. Dalian nvxing Aizibing ganranzhe zengduo (Increasing Rate of HIV Infected Women in Dalian). In *Zhongguo Funv Bao* (*Chinese Women's Newspaper*), 3. Beijing: National Federation of Women.

Sigley, Gary. 2001. Keep it in the family: Government, marriage, and sex in contemporary China. In *Borders of being: Citizenship, fertility, and sexuality in Asia and the Pacific*, ed. Margaret Jolly and Kalpana Ram, 118–53. Ann Arbor: University of Michigan Press.

Silva, Martha, Dominique Meekers, and Megan Klein. 2003. Determinants of condom use among youth in Madagascar. PSC Research Report No. 55. PSI (Population Services International). Washington, DC.

Singer, Merrill. 1994. AIDS and the health crisis of the U.S. urban poor: The perspective of critical medical anthropology. *Social Science & Medicine* 39 (7): 931–48.

———. 1998. *The political economy of AIDS.* Amityville, NY: Baywood.

Singer, Merrill, Candida Flores, Lani Davison, Georgine Burke, Zaida Castillo, Kelley Scalon, and Migdalia Rivera. 1990. SIDS: The economic, social, and cultural context of AIDS among Latinos. *Medical Anthropology Quarterly* 4 (1): 72–114.

Singer, Merrill, Zhongke Jia, Jean J. Schensul, Margare Weeks, and J. Bryan Page. 1992. AIDS and the IV drug user: The local context in prevention efforts. *Medical Anthropology* 14:285–306.

Singhal, Arvind, and Everett M. Rogers. 1999. *Entertainment-education: A communication strategy for social change.* Mahwah, NJ: Lawrence Erlbaum.

Singhal, Arvind, and Everett Rogers. 1988. Television soap operas for development in India. *Gazette* 41:109–26.

Smith, Beverly. 1996. AIDS: Religion and medicine in rural Kenya. In *AIDS education: Interventions in multicultural societies*, ed. Inon I. Schenker, Galia Sabar-Friedman, and Francisco S. Sy, 239–49. New York: Plenum.

Smyth, Fiona. 1998. Cultural constraints on the delivery of HIV/AIDS prevention in Ireland. *Social Science & Medicine* 46 (6): 661–72.

Sobo, E. J. 1993. Inner-city women and AIDS: Psychosocial benefits of unsafe sex. *Cultural Medical Psychiatry* 17:454–85.

———. 1994. Attitudes towards HIV testing among impoverished urban African-American women. *Medical Anthropology* 16:1–22.

———. 1995a. *Choosing unsafe sex: AIDS-risk denial among disadvantaged women*. Philadelphia: University of Pennsylvania Press.

———. 1995b. Finance, romance, social support, and condom use among impoverished inner-city women. *Human Organization* 54:115–28.

———. 1998. Love, jealousy, and unsafe sex among inner-city women. In *The political economy of AIDS*, ed. Merrill Singer, 75–103. Amityville, NY: Baywood.

Solinger, Dorothy. 1999. *Contesting citizenship*. Berkeley: University of California Press.

Sontag, Susan. 1989. *AIDS and its metaphors*. Trans. Michael Henry Heim. New York: Farrar, Straus, and Giroux.

Stine, Gerald. 2007. *AIDS update 2007*. New York: Pearson.

Stoler, Ann Laura. 1991. Carnal Knowledge and imperial power: Gender, race, and morality in colonial Asia. In *Gender at the crossroads of knowledge*, ed. M. de Leonardo. Berkeley: University of California Press.

Strand, David. 1989. *Rickshaw Beijing: City people and politics in the 1920s*. Berkeley: University of California Press.

Sun, Minghua. 2007. Dalian Aizibing yiqing shangsheng qushi mingxian 7 cheng chengwei xingchuanbo (Dalian AIDS infection rate is rapidly rising, more than 70% from sexual transmission). In *Zhong guang wang*. Beijing: Zhongguo Guangbowo (Chinese Broadcast Net).

T., F. 1977. Fertility control and public health in rural China: Unpublicized problems. *Population and Development Review* 3 (4): 482–85.

Taylor, Christopher C. 1990a. AIDS and the pathogenesis of metaphor. In *Culture and AIDS: The human factor*, ed. D. Feldman, 55–65. Westport, CT: Praeger.

———. 1990b. Condoms and cosmology: The "fractal" person and sexual risk in Rwanda. *Social Science & Medicine* 31 (9): 1023–28.

Thuy, Nguyen, Thi Thanh, Christina P. Lindan, Nguyen Hoan, John Barclay, and Khiem Ha. 2000. Sexual risk behavior of women in entertainment services. *AIDS and Behavior* 4:93–101.

Tien, H. Yuan. 1965. Sterilization, oral contraception, and population control in China. *Population Studies* 18 (3): 215–35.

———. 1973. *China's population struggle: Demographic decisions of the People's Republic, 1949–1969*. Columbus: Ohio State University Press.

Tierney, John. 1990. With "social marketing," condoms combat AIDS. Special to the *New York Times*, September 18.

Tong, Wei, and Dejun Li. 1996. Dalian shouci faxian HIV Ganranzhe (The first HIV infected people in Dalian). *Zhongguo Xingbing Aizibing Fangzhi* (*Prevention of STI and AIDS in China*) 4:15–16.

Tong, Wei, and Xiaoming Yang. 1998. Dalianshi shoulei jingwai yiyuanxing ganren HIV baogao [The first case of overseas origin of HIV infection in Dalian]. *Zhongguo Xingbing Aizibing Fangzhi* (*Prevention of STI and AIDS in China*) 3:103.

Tran, Trung Nam, Roger Detels, and Hoang Phuong Lan. 2006. Condom use and its correlates among female sex workers in Hanoi, Vietnam. *AIDS and Behavior* 10 (2): 159–67.

Travers, Michele, and Lydia Bennett, ed. 1996. *AIDS, women, and power*. New York: Taylor & Francis.

Triechler, Paula. 1987. AIDS, homophobia, and biomedical discourse: An epidemic of signification. *Cultural Studies* 1 (3): 263–305.

Tu, Xiao. 2001. Wohe Aizibingren youge yuehui (I have an appointment with an AIDS patient). *Jia Ting (Family)* 1 (251): 20–24.

UNAIDS. 2003. Join the fight against AIDS in China. Geneva, Switzerland: Joint United Nations Programme on HIV/AIDS (UNAIDS).

Vance, Carol. 1982. Pleasure and danger: Toward a politics of sexuality. In *Pleasure and danger: Exploring female sexuality*, ed. Carol S. Vance. London: Pandora.

Van den Hoek A., Y. L. Fu, N. H. T. M. Dukers, Z. H. Chen, J. T. Feng, L. N. Zhang, and X. X. Zhang. 2001. High prevalence of syphilis and other sexually transmitted diseases among sex workers in China: Potential for fast spread of HIV. *AIDS* 15 (6): 753–59.

Van den Hoek, J., Roel A. Coutinho, Harry J. A. van Haastrecht, Alt W. van Zadel-hoff, and Jaap Goudsmit. 1988. Prevalence and risk factors of HIV infections among drug users and drug using prostitutes in Amsterdam. *AIDS* 2 (1): 55–60.

Van Gulik, R. H. 2003. *Sexual life in ancient China*. Trans. Li Ling and Guo Xiaohui. New York: Brill Academic.

Vanwesenbeeck, Ine, Ron De Graaf, Gertjan Van Zessen, Cees J. Starver, Jan H. Visser. 1993. Protection styles of prostitutes' clients: Intention, behavior, and considerations in relation to AIDS. *Journal of Sex Education and Therapy* 19:79–92.

Vaughan, Peter W., and Everett M. Rogers. 2000. A staged model of communication effects: Evidence from an entertainment-education radio soap opera in Tanzania. *Journal of Health Communication* 5:203–27.

Waddell, Charles E. 1996a. HIV and the social world of female commercial sex workers. *Medical Anthropology Quarterly* 10:75–82.

———. 1996b. Female sex work, non-work sex and HIV in Perth. *Australian Journal of Social Issues* 31 (4): 410–24.

Waldby, Cathy, Susan Kippax, and June Crawford. 1993. Cordon sanitaire: "Clean" and "unclean" women in the AIDS discourse of young men. In *AIDS: Facing the second decade*, ed. Peter Davies Peter Aggleton and Graham Hart, 29–39. London: Falmer Press.

Walden, Vivien Margaert, Kondwani Mwangulube, and Paul Makhumula-Nkhoma. 1999. Measuring the impact of a behaviour change intervention for commercial sex workers and their potential clients in Malawi. *Health Education Research* 14 (4): 545–54.

Walters, Ian. 2004. Dutiful daughters and temporary wives: Economic dependency on commercial sex in Vietnam. In *Sexual cultures in East Asia: The social construction of sexuality and sexual risk in a time of AIDS*, ed. Evelyne Micollier, 76–97. London: RoutledgeCurzon.

Wang, Aili. 1995. Dangjin chengshi hunyin yanbian de jiben biaozheng jiqi zouxiang (The basic characteristics and trend of current city marriage changes). *Women of China* 2:20–21.

Wang, Changshan. 2005. Jianjue ezhi aizibing de liuxing he manyan (Insist in blocking the spread of AIDS). *Nanfang Zhoumo* (*South Weekend*), June 23:1.

Wang, Gan. 1999. Conspicuous consumption, business networks, and state power in a Chinese City. PhD diss., Yale University.

Wang, Jichuan, Baofa Jiang, Harvey Siegal, Russel Falck, and Robert Carlson. 2001a. Sexual behavior and condom use among patients with sexually transmitted diseases in Jinan, China. *American Journal of Public Health* 91 (4): 650–51.

———. 2001b. Level of AIDS and HIV knowledge and sexual practices among sexually transmitted disease patients in China. *Sexually Transmitted Disease* 28 (3): 171–75.

Wang, Jinling. 2006. *Zhongguo funu fazhan baogao No. 1* (*A report on Chinese women's development*). Beijing: Shehui Kexue Wenxian Chubanshe (Social Science Publishing House).

Wang, Lijuan. 2007. Condom promotion for HIV/AIDS prevention in China: The role of epistemic community. Eighth International Congress on AIDS in Asia and the Pacific, Colombo August 19–23.

Wang, Longde. 2007. Overview of the HIV/AIDS epidemic, scientific research and government responses in China. *AIDS* 21 (8): 3–7.

Wang, Shancheng. 1956. Tan xingshenghuo (Talk about sex life). *Women of China* 8:30.

Wang, Wei. 2005. Dalian ¼ yihun nuxing fanghuan biyun. In *Bandao Chenbao* (*Bandao Morning Newspaper*), June 13.

Wang, Wenbin, Zhiyi Zhao, and Mingxun Tan. 1956. *Xing de zhishi* (*Sexual knowledge*). Beijing: Renmin Weisheng Chubanshe (People's Hygiene Publishing House).

———. 1957. *Xing de zhishi* (*The knowledge of sex*). Beijing: Kexue Puji Chubanshe.

Wang, Yan. 2007. Anti-HIV/AIDS prevention within sex workers in Sichuan. Seminar on HIV Prevention and Sex Work, UNDP, Beijing. June 4.

Wang, Yuefeng. 2005. Qiantan weifa fanzui renyuanzhong ganran aizibingren de guanli (On the management of AIDS Criminals). In *Zhongguo Aizibing fangzhi* (*HIV Prevention in China*). Beijing: Capital University of Medical Science.

Wang, Zhenhua. 2003. Anquandai Yu anquantao (Safety belt and condoms). *Shandong Qingnian Bao* (*Shandong Youth Newspaper*), February 22.

Warr, Deborah J., and Priscilla M. Pyett. 1999. Difficult relations: Sex work, love, and intimacy. *Sociology of Health and Illness* 21:290–309.

Wawer, Maria J., Chai Podhisita, Uraiwan Kanungsukkasem, A. Pramualratana, and R. Mcnamara. 1996. Origins and working conditions of female sex workers in urban Thailand: Consequences of social context for HIV transmission. *Social Science and Medicine* 42:453–62.

Webb, Douglas. 1997. *HIV and AIDS in Africa*. Cape Town: David Philip.

Wei, Jingjing. 2002. "Yetibiyuntao" nengfou taolao jiankang he kuaile (Can liquid condom offer health and happiness). *Zhongguo Funu* (*Women of China*) 12 (614): 6.

Wei, Ping. 2005. Nuxingjiankang yao baowo sange zhuanxingqi (Caring female health should grasp three transitional period). *Jiankang Bao* (*Health News*) 7:7.

Weissman, Carol S., Constance A. Nathanson, Margaret Ensminger, Martha A. Teitelbaum, J. Courtland Robinson, and Stacey Plichta. 1989. AIDS knowledge, perceived risk, and prevention among adolescents of a family planning clinics. *Family Planning Perspectives* 21:213–17.

Wen, Chihua. 2002. No condoms, please, we're Chinese men. *Asia Times Online*, April 11.

Wen, Jin. 1958. Jieyu xuanchuan zai yuhuasha chang (Publicity of birth control in Yuhua Yarn Factory). *Women of China* 6:30–31.

Wen, Li. 2005a. Nuren, diode de zuihou shouhuzhe: Jiatingzhulian, dizhifubaizhufangxian (Women are the last defenders of ethics: Helping with the cleanness of the family, resisting corruption). *Zhongguo Funu* (*Women of China*), September 2:10–11.

———. 2005b. Nuren, diode de zuihou shouhuzhe: Jiatingzhulian, dizhifubaizhufangxian (Women are the last defenders of ethics: Helping with the cleanness of the family, resisting corruption). *Zhongguo Funu* (*Women of China*), October (2): 51–52.

———. 2005c. Nuren, diode de zuihou shouhuzhe (Women are the last defenders of ethics). *Zhongguo Funu* (*Women of China*), July 2:52–53.

———. 2005d. Nuren, diode de zuihou shouhuzhe (Women are the last defenders of ethics). *Zhongguo Funu* (*Women of China*) no. 655 (August 2): 52–53.

Wen, Zhenxiu. 2002. Social marketing to high risk groups in Yunnan and Sichuan, China. International Conference on AIDS, Manchester July 7–12.

Werner, David. 1977. Where there is no doctor: A village health care handbook. Palo Alto, CA: Hesperian Foundation.

Wight, Daniel. 1994. "Boys" thoughts and talk in a working class locality of Glasgow. *Sociological Review* 42:703–37.

Williams, Sophie, and Lenore Lyons. 2008. It's about bang for your buck, bro: Singaporean men's online conversations about sex in Batam, Indonesia. *Asian Studies Review* 32 (March): 77–97.

Wilson, David, Babusi Sibanda, Lilian Mboyi, Sheila Msimanga, and Godwin Dube. 1990. A pilot study for an HIV prevention programme among commercial sex workers in Bulawayo, Zimbabwe. *Social Science & Medicine* 31:609–18.

Wilton, Tamsin, and Peter Aggleton. 1991. Condoms, coercion, and control: Heterosexuality and the limits to HIV/AIDS education. In *AIDS: Responses, interventions, and care*, ed. Peter Aggelton, Peter Davies, and Graham Hart, 149–56. London: Falmer Press.

Wiutehead, Tony L. 1997. Urban low-income African American men, HIV/AIDS, and gender identity. *Medical Anthropology Quarterly* 11 (4): 411–47.

Wojcicki, Janet, and Josephine Malala. 2001. Condom use, power, and HIV/AIDS risk: Sex-workers bargain for survival in Hillbrow/Joubert Park/Berea, Johannesburg. *Social Science & Medicine* 53:99–121.

Wolf, Margery. 1972. *Women and the family in rural Taiwan*. Stanford, CA: Stanford University Press.

Wong, M., I. Lubek, B. C. Dy, S. Pen, S. Kros, and M. Chhit. 2003. Social and behavioural factors associated with condom use among direct social workers in Siem Reap, Cambodia. *Sexually Transmitted Infections* 79:163–65.

Worth, Dooley. 1989. Sexual decision-making and AIDS: Why condom promotion among vulnerable women is likely to fail. *Studies in Family Planning* 20 (6): 297–307.

Wu, Bo. 2004. Jujue xingbing de fangshenshu (STI Prevention). *Baojian yu Shenghuo* (health and life) 7:38.

Wu, Yin. 1956. Zaibuneng youyi buding le (No longer hesitant). *Women of China* 7:26.

Wu, Zunyou, Keming Rou, Manhong Jia, Song Duan, and Sheena Sullivan. 2007. The first community-based sexually transmitted disease/HIV intervention trial for female sex workers in China. *AIDS* 21 (8): 89–94.

Wu, Zunyou, Rou Keming, and Cui Haixia. 2004. The HIV/AIDS epidemic in China: History, current strategies, and future challenges. *AIDS Education and Prevention* 16 (Suppl. A): 7–17.

Wulfert, Edelgard, and Choi K. Wan. 1995. Safer sex intentions and condom use viewed from a health belief, reasoned action, and social cognitive perspective. *Journal of Sex Research* 32 (4): 299–312.

Xi, Li. 2005. Liaojie AIDS yufang xinguannian (New views of prevention of AIDS). *Jiankang Wenzhai Bao* (*Health and Digest Newspaper*) no. 700 (February 20): 8.

Xi, Zhu. 1964. Airen de sixiang kaile qiao (My husband's thought finally got straightened out). *Women of China* 8:30.

Xiao, Hua. 1965. Xinhun neng biyun ma (Can we proceed contraception as a newly married couple). *Women of China* 12:31.

Xiao, Liu. 2001. Siwang yinying longzhao "Xiaojie" qunti (Hostesses are shadowed by Death). *Dalian Wanbao* (*Dalian Evening Newspaper*), 18.

Xie, Hanping. 1997. Buzhengchang de xingshenghuo weihaiduo (More harms of abnormal sex life). *Women of China* (*Zhongguo Funu*) 3:55.

Xin, Ren. 1999. Prostitution and economic modernization of China. *Violence against Women* 5 (12): 1411–36.

Xiwen, Zheng, Qu Shuquan, Kevin Yiee, and Jeffrey Mandel. 2000. HIV risk among patients attending sexually transmitted disease clinics in China. *AIDS and Behavior* 4 (1).

Xu, Zongxiu. 1964. Zai jieshou le liangci jiaoxun yihou (After receiving two lessons). *Women of China* 5:30.

Yang, Chun. 2004a. He aizibingdu saipao (Racing AIDS virus). *Qingnian Shixun* (*Youth Express*), June 1 (207): 1.

———. 2004b. He aizibingdu saipao (Racing AIDS virus). *Qingnian Shixun* (*Youth Express*), no. 207 (July 22): 1.

Yang, Junlan, and Yonggang Li. 1999. Liaoningsheng 1998 nian xingbing liuxing qingkuang ji fenxi (An epidemiological analysis of STI situation in Liaoning Province). *Zhongguo Xingbing Aizibing Fangzhi* (*Prevention of STI and AIDS in China*) 5 (6): 243–45.

Yang, Junlan, and Shibo Zhang. 1997. Liaoningsheng xingbing yiqing fenxi (An analysis of STI situation in Liaoning Province). *Zhongguo Xingbing Aizibing Fangzhi* (*Prevention of STI and AIDS in China*) 3 (6): 241–42.

Yang, Mayfair Mei-hui. 1999a. From gender erasure to gender difference: State feminism, consumer sexuality, and women's public sphere in China. In *Spaces of their own: Women's public sphere in transnational China*, ed. Mayfair Mei-hui Young, 35–67. Minneapolis: University of Minnesota Press.

———. 1999b. Introduction. In *Spaces of their own: Women's public sphere in transnational China*. ed. Mayfair Mei-hui Yang, 1–31. Minneapolis: University of Minnesota Press.

Yao, Ge. 2002. *Xinhun weisheng quanshu* (*Hygiene for newly married couple*). Fuzhou: Fujian Kexue Jishu Chubanshe.

Yeakley, Anna M., and Larry M. Gant. 1997. Cultural factors and program implications: HIV/AIDS interventions and condom use among Latinos. *Journal of Multicultural Social Work* 6:47–71.

Yi, Ming. 2001. Anquantao chulu hezai (Where is the outlet for condom). *Keji Ribao* (*Science Daily*), December 30.

Yin, Chengxi. 2006. Bieyaomohua Chongqingde anquantao xiangmu (Don't Demonize Chongqing's condom program). *Xiandai Kuaibao* (*Modern Express Newspaper*), September 7.

Yip, Ray. 2006. Opportunity for effective prevention of AIDS in China: The strategy of preventing secondary transmission of HIV. In *AIDS and Social Policy in China*, ed. Arthur Kleinman, Joan Kaufman, and Tony Saich, 177–89. Cambridge, MA: Harvard University Asia Center.

You, Qing. 2000. Dangdai nudaxuesheng xingxingwei xintailu (A record of modern female university students' sex conduct and psyche). *Changjiang Ribao* (*Yangtsze River Daily*), October 26:4.

Yu, Dingzhen. 1957. Wode tongku cong helai (Where does my pain come from). *Women of China* 3:14–15.

Yu, Guangyuan. 1956. *Xing zhishi* (*Knowledge of sex*). Shanghai: Shanghai Weisheng Chubanshe.

Yu, Hushi. 2002. Pinglun: Woguo anquantao guanggao jiejin xiaoxiao de yige jinbu (Comments: A small progress of condom ads). *Fazhi Ribao* (*Law Daily*), December 3.

Yu, Muxia. 1935. Shanghai linzhao (Shanghai tidbits). Shanghai: Shanghai hubaoguan chubanbu (Shanghai Hubaoguan Publishing House).

Yu, Ping. 1957. Ku nao (Frustration). *Women of China* 3:14–15.

Yu, Sheng. 1958. Biyun geiwo dailai de haochu (Birth control brought me advantages). *Women of China* 4:31.

Yu, Xian. 1965. Fangle biyun huan buhui yingxiang shenti jiankang (IUD will not affect physical health). *Women of China* 10:31.

Yu, Ying. 2003. *Bi yun (Contraception)*. Haikou: Nanhai Chuban Gongsi.

Yuan, Feng, and Jie Yun. 1990. Baohuzhencao de mijue (Secret to chastity). *Cha yu Fan Hou (After Tea and Meals)* 5:6.

Zeng, Liming. 2002. Zhongguo yanzhi chu neng shamie duozhong bingdu de yeti biyuntao (China has invented liquid condoms that can kill a plethora of viruses). *Beijing Qingnian Bao (Beijing Youth Newspaper)*, October 18.

Zhang, Baichuan, D. Liu, X. Li, and T. Hu. 2001. A survey on HIV/AIDS related high risk behaviors and affecting factors of men who have sex with men in Mainland China. *Chinese Journal of Sexually Transmitted Infections* 1 (1): 7–16.

Zhang, Li. 2001. *Strangers in the city: Reconfigurations of space, power, and social networks within China's floating population*. Stanford, CA: Stanford University Press.

Zhang, Lizhi. 1966. Dui shehuizhuyi youli de shi won eng bu dai tou ma (Can't I lead others to help with socialism?) *Women of China* 1:24.

Zhang, L. Y., X. Gao, Z. W. Dong, Y. P. Tan, Z. L. Wu. 2002. Premarital sexual activities among students in a university in Beijing, China. *Sexually Transmitted Disease* 29 (4): 212–15.

Zhang, Qing. 2004. Nuxing ganranzhe bili you jiaoda fudu shangsheng (Rising ratio of the female infected). *Zhongguo Funubao (Chinese Women Newspaper)*, December 6:5.

Zhang, Wei. 2006. Anquantao de zhunru yiyi (The meaning of entry of condoms). *Beijing Yule Xinbao (Beijing Entertainment Newspaper)*, February 28.

Zhang, Youfang, and Lei Feng. 2001. *Abortion*. Beijing: Renmin Weisheng Chubanshe.

Zhang, Yujie. 1964. Yici meiyou jing zuzhi de zuotanhui (An unorganized seminar). *Women of China* 5:30–31.

Zhang, Zhongyuan, and Xun An. 2002. *Shengzhi jiankang zhinan (The handbook of biological health)*. Chengdu: Sichuan Renmin Chubanshe.

Zhao, Jian. 2004. Yufang AIDS xuanchuan yu ganga (Embarrassment of advocating AIDS prevention). *Beijing Qingnian Bao (Beijing Youth Newspaper)*, November 30.

Zhao, Pengfei. 2005. *Female condom consultation in China*. Summer Report of Meeting, Beijing: The Female Health Foundation. April 26.

Zhao, Ting. 2002. Zhangfu chuci "yuegui" shi (The first time when the husband went astray). *Zhongguo Funu (Women of China)* 6 (2): 32.

Zhao, Xueyi. 2006. Dalian "Fang Ai" zhaoer tingduo (Dalian has many AIDS-prevention means). In *Liaoning Fazhi Bao (Liaoning Law Newspaper)*. Shenyang.

Zhao, Zhiyi, and Hongzhao Song. 1955. *Biyun changshi (Common knowledge about contraception)*. Beijing: renmin weisheng chubanshe (People's Hygiene Publishing House).

Zhao, Zhongwei. 2006. Towards a better understanding of past fertility regimes: The ideas and practices of controlling family size in Chinese history. *Continuity and Change* 21 (1): 9–35.

Zhen, Yulan, and Longyin Mao. 2003. Chengshilide xin zhencao yundong (New Chastity Campaign in the City). *Shenzhen Qingnian (Shenzhen Youth)* 5 (174): 54–55.

Zheng, Lingqiao. 2004. Huanqi minzhong shi fangkong AIDS de dunpai (Mobilizing the populace is the shield against AIDS). *Jiankang Bao (Health Newspaper)*, March 11:2.

Zheng, Tiantian. 2003. Consumption, body image, and rural-urban apartheid in contemporary China. *City and Society* 15 (2): 143–63.

———. 2004. From peasant women to bar hostesses: Gender and modernity in post-Mao Dalian. In *On the move: Women and rural to urban migration in contemporary China*, ed. Arianne Gaetano and Tamara Jacka, 80–108. New York: Columbia University Press.

———. 2006. Cool masculinity: Male clients' sex consumption and business alliance in urban China's sex industry. *Journal of Contemporary China* 15 (46): 161–82.

———. 2007. Claim for an equal social status: An ethnography of China's sex industry. In *Working in China: Ethnographies of labor and workplace transformation*, ed. Ching Kwan Lee, 124–44. New York: Routledge.

———. 2008a. Commodifying romance and searching for love: Rural migrant bar hostesses' moral vision in post-Mao Dalian. *Modern China* 35 (4): 442–76.

———. 2008b. Complexity of life and resistance: Informal networks of rural migrant karaoke bar hostesses in urban Chinese sex industry. *China: An International Journal* 6 (1): 69–95.

———. 2009. *Red lights: The lives of sex workers in postsocialist China*. Minneapolis: University of Minnesota Press.

Zheng, Xiwen, Kyung-Hee Choi, Xiwen Zheng, Shuquan Qu, Kevin Yiee, Jeffrey Mandel. 2000. HIV risk among patients attending sexually transmitted disease clinics in China. *AIDS Behavior* 4 (1): 111–19.

Zhong, Xueping. 2000. *Masculinity besieged? Issues of modernity and male subjectivity in Chinese literature of the late twentieth century*. Durham, NC: Duke University Press.

Zhou, Efen. 1955. Biyun wenti daduzhe wen (Answer to the readers' questions on contraception). *Women of China* 12 (74): 26.

———. 1956. Tan zhongyao biyun (About contraceptive in Chinese medicine). *Women of China* 9:30–31.

Zhou, Huadong. 2005. Sancheng renliu nuxing wei zaixiao daxuesheng (Thirty percent of the women going through abortion are female college students). *Bandao Chenbao*, June 15, A13.

Zhu, Junlun, and Xiansong Li. 1993. *Xinhun bidu, fuqi biben (A must read for newly married couples and spouses)*. Chengdu: Chengdu Keji Daxue chubanshe.

Zhu, Kun. 2002. Dang xing zaoyu aizi shashou (When sex encounters the killer of AIDS). *Xin Zhou Kan (New Weekly)*, February 19.

Index